Signal Processing with Lapped Transforms

For a complete listing of the *Artech House Telecommunications Library,*
turn to the back of this book . . .

Signal Processing with Lapped Transforms

Henrique S. Malvar
Universidade de Brasília, Brazil

ARTECH HOUSE

Boston • London

Library of Congress Cataloging-in-Publication Data

Malvar, Henrique S., 1957-
  Signal processing with lapped transforms / Henrique S. Malvar.
    p. cm.
  Includes bibliographical references and index.
  ISBN 0-89006-467-9
  1. Signal processing–Digital techniques. 2. Signal processing–
Mathematics. 3. Transformations (Mathematics) I. Title.
II. Title: Lapped transforms. III. Series.
TK5102.2M275 1991   91-35984
621.382'2–dc20   CIP

British Library Cataloguing in Publication Data

Malvar, Henrique S.
  Signal processing with lapped transforms.
  I. Title.
  621.3822

ISBN 089006-467-9

© 1992 ARTECH HOUSE, INC.
685 Canton Street
Norwood, MA 02062

International Standard Book Number: 0-89006-467-9
Library of Congress Catalog Card Number: 91-35984

10 9 8 7 6 5 4 3 2 1

To my parents, Henrique and Gilséa

Contents

Preface

Digital signal processing (DSP) has been a growing field for more than three decades. With the availability of fast integrated circuits dedicated specifically to DSP applications, we now live in a world where DSP is not just a hot research topic, but part of our everyday life. If we look at what is attached to a standard telephone line in our modern office, for example, we see modems, fax machines, and tapeless answering machines, all of which could not exist without DSP. We could certainly spend many pages describing examples of DSP applications.

Advances in DSP have been so many that specialized areas within it are themselves becoming new fields. Among them, we may cite speech processing, image processing, adaptive filtering, and multirate signal processing. In all of these areas, fast transforms are frequently used, because it is often more efficient to process a signal in the transform domain.

The purpose of this book is to present to the reader a complete reference on the theory and applications of a new family of transforms, called *lapped transforms* (LTs). These transforms can be used in many applications, such as filtering, coding, spectral estimation, and any others where a traditional block transform is employed, such as the discrete Fourier transform (DFT) or the discrete cosine transform (DCT). In many cases, LTs will lead to a better complexity versus performance trade-off than other transforms. Until now, the theory of lapped transforms and many of their applications have appeared in theses, journal articles, and conference proceedings. This is the first book in which all of the known results are put together in an organized form.

We believe that this book is a useful reference for design engineers, graduate students, and researchers involved with DSP applications that make use of fast transforms. Many signal processing systems employing fast transforms are presented, as well as evaluations of implementations of those systems. Thus, the reader with a practical application in mind will be able to put LTs to work to his or her benefit immediately. For that purpose, we have included in the appendices listings of computer programs with fast algorithms for lapped transforms, as well as programs for traditional block transforms based on optimized algorithms.

As is the case with any new topic, there are many interesting theoretical issues involving LTs, for example, the relationships that exist among LTs, multirate filter banks, and discrete wavelet transforms. Throughout the book, there are sections devoted entirely to these and other theoretical aspects. The reader is only assumed to have a solid background in the basic theory of discrete-time signal processing, including the fundamentals of random signal representation and introductory linear algebra. Undoubtedly, this book will be even more useful to those already familiar with the implementation of signal processing systems that employ fast transforms or filter banks.

Chapter 1 starts with a brief review of signal models and the basic definitions and properties of traditional block transforms and lapped transforms. A brief history of the development of lapped transforms is also presented. In Chapter 2, some of the applications of block transforms are discussed, with emphasis on the DFT, DHT, and the DCT. The current state-of-art of fast algorithms for these transforms is reviewed, and the best known procedures for their computation are presented.

The basics of multirate signal processing are discussed in Chapter 3, with the purpose of studying maximally-decimated filter banks, which are essential building blocks of subband coding systems. Special attention is given to quadrature mirror filter (QMF) banks and perfect reconstruction (PR) filter banks, including conjugate quadrature filters (CQF). The chapter discusses the fundamental idea that transform coding is in fact a special case of subband coding, and also discusses briefly the applications of subband signal processing.

In Chapter 4, the theory and properties of the lapped orthogonal transform (LOT) are studied in detail. The theoretical aspects leading to the PR property of lapped transforms are discussed, within the context that a lapped transform is a natural extension of a regular block transform. This extension is directed towards turning the transform into a filter bank with improved frequency resolution. Design techniques and fast algorithms for the LOT are presented. The coding performance of the LOT, which is better than that of the DCT, is also discussed.

The modulated lapped transform (MLT) family of LTs is studied in Chapter 5, together with its generalized version, the extended lapped transform (ELT). A detailed discussion of the design techniques for the generation of optimized MLTs and ELTs is presented. Fast algorithms for the MLT and ELT are also presented at a level of detail previously unavailable in the literature. The chapter ends with a theoretical discussion of the coding performance of the MLT and ELT.

The hierarchical lapped transform (HLT) is discussed in Chapter 6. HLTs are useful for multiresolution signal analysis and coding because HLTs are in fact filter banks with subbands of unequal widths, and impulse responses of different lengths. Tree structures for the HLT are discussed, as well as the connections with the discrete wavelet transform.

Applications of lapped transforms are discussed in Chapter 7. Examples of the use of LTs in signal filtering are considered, with emphasis on adaptive filters and variable filters. The use of LTs in signal coding is also discussed, with many examples of the results obtained with speech and image coders based on LTs. From these results, it becomes clear that one of the main advantages of lapped transforms over traditional block transforms is the strong reduction in the discontinuities in the reconstructed signal at the block boundaries, the so-called *blocking effects*.

The appendices present valuable information for the reader interested in putting the ideas in this book to work immediately in his (or her) application that requires a block transform or a filter bank. When the desired number of bands is two, a good alternative for the implementation of perfect-reconstruction filter banks is the conjugate quadrature filters (CQFs). Thus, we have included a table of CQF coefficients in Appendix A. Computer programs for fast computation of some of the most commonly used block transforms are presented in Appendix B. In Appendix C there are several tables of butterfly angles for the MLT, and in Appendix D computer programs for fast computation of LTs are presented. The programs are all written in the "C" language for increased portability.

Acknowledgements

There are many people who had a strong influence on the material presented in this book, and to whom I would like to thank. Professor David H. Staelin, my thesis supervisor at M.I.T. and a good friend, has always been very supportive and encouraging, giving me the right advice on everything. He was the originator of the term *lapped transform*. The research team that developed the basic ideas behind the lapped orthogonal transform at M.I.T., back in 1984, also included Philippe Cassereau, Brian Hinman, and Jeff Bernstein. Many encouraging discussions on the theory and applications of lapped transforms and related topics were held with Dr. A. Briançon, C. Clapp, M. Cruvinel, Dr. G. de Jager, Prof. P. S. R. Diniz, R. Duarte, Dr. P. Duhamel, Dr. S. Ericsson, Dr. D. Le Gall, Dr. F. A. O. Nascimento, A. Popat, Prof. K. R. Rao, R. Saraiva Jr., Dr. J. Shapiro, Prof. M. J. T. Smith, Dr. B. Szabo, Dr. A. Tabatabai, Prof. P. P. Vaidyanathan, Prof. M. Vetterli, Prof. A. S. Willsky, and Dr. G. Wornell.

In particular, I would like to express my sincere thanks to Ricardo L. de Queiroz for his many suggestions on the manuscript, and for carrying out all of the computer simulations of the applications of lapped transforms to image processing. I am also thankful to Eduardo M. Rubino, for writing a family of TeX[1] device drivers that allowed this book to be entirely typeset by the author. Eduardo also wrote the software that produced the half-tone images of Chapter 7. The encouragement that I received from Mark Walsh, Pamela Ahl, Kim Field, and Dennis Ricci of Artech House helped keep my peace of mind as I was writing this book. It was certainly a pleasure working with them.

The financial support from the Brazilian Government, through the *Conselho Nacional de Desenvolvimento Científico e Tecnológico – CNPq*, is gratefully acknowledged. CNPq supported most of my research on lapped transforms since 1987, through grants nos. 404.963-87, 404.519-88, 300.159-90, and 600.047-90.

Finally, a special note of gratitude goes to my wife Regina Helena, my daughter Ana Beatriz, and my son Henrique. Each hour spent writing this book was an hour taken away from them; and there were many, many such hours. Without their patience and understanding, this book could not have been written.

---

[1]TeX is a trademark of the American Mathematical Society.

Chapter 1

Introduction

This is a book on signal processing with lapped transforms (LTs). At first, this might seem to be an obscure subject because LTs are relatively new. This is not so, however, and LTs are becoming more attractive for a wide variety of applications. This is mainly because LTs are a special family of filter banks that can be easily designed and implemented, even for a large number of subbands. Throughout this book, we shall see that LTs can lead to better system performance than other more usual transforms, like the discrete Fourier transform (DFT) or discrete cosine transform (DCT), in applications such as image coding, speech coding, and adaptive filtering. In any application where a block transform or a filter bank is employed, a lapped transform can also be used, since block transforms and lapped transforms can always be viewed as special cases of filter banks. As we will see in later chapters, in many cases LTs will lead to a better signal representation or reduced computational complexity, or both, when compared to the most commonly used block transforms or filter banks.

Before we start discussing LTs, it is important that we review the basics of discrete-time signal representation, so that we make clear what we mean by a signal. This is what we shall do in Section 1.1, where we will follow a statistical approach toward signal modeling. Looking at signals as sample functions of stochastic processes helps to predict quite accurately average system performance. We must also review the basic concepts behind traditional block transforms to support our later discussion of LTs, and this is the goal of Section 1.2. A brief introduction to lapped transforms, including a review of their history, is presented in Section 1.3.

1.1   Signal Models

We call signal a sequence of real numbers that carry some meaningful information. Such a sequence of numbers generally comes from periodic sampling of a physical waveform, as is the case in speech and image processing. However, it can also represent an economic time series, such as the daily Dow Jones index, for example. The length of the signal can be finite or not, and that makes little difference in most situations. When we refer to signals as sequences of numbers, we are already considering them as functions of discrete time (or discrete space, according to the independent variable) [1], and so we will not consider throughout this book signals that are continuous waveforms. This is nowadays a natural trend in signal processing; whenever we can process a signal digitally, it is generally cheaper and more efficient to do so than to use analog signal processing.

In order to process signals digitally, we need not only discrete time but also discrete amplitude, because we must allocate a finite number of bits of storage to represent a number in a digital computer. However, as usual in most texts dealing with digital signal processing, we will always assume that the signal amplitudes are continuous. Thus, each signal sample is represented by a real number. Whenever appropriate, we will discuss briefly the effects of discretizing the signal amplitudes. In most cases we will simply assume that the errors in the signal representation due to finite-precision storage are small enough to be negligible.

1.1.1   Deterministic and Stochastic Models

Depending on the application, signals can be viewed either as deterministic or as random waveforms. We say that a signal is deterministic when it is specified by a table of values or by some mathematical function that defines the values of each sample (some parameters of the function may be unknown). An example of a deterministic waveform is a sinusoid, as shown in Fig. 1.1, which may be generated by a modem in a digital communication system. For a random signal, we may not know its precise waveform in a particular observation of it, but we have a probabilistic model or measured statistics that describe some properties of the signal, such as mean, variance, and autocorrelation.

Random signals are good models for real-world signals such as speech and images [2]. Two examples of random signals are also shown in Fig. 1.1, where we have a typical sample waveform of a white noise process (in which all frequencies

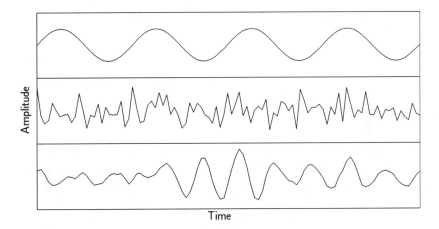

Figure 1.1. Examples of signals. Top: deterministic; middle: random, white; bottom: random, colored. Note that these are discrete-time signals; the points corresponding to sample values were connected by straight lines for easier viewing.

from 0 to π appear with equal contribution), and also a sample waveform of a colored random signal (in which some frequency components have higher energies than others).

A deterministic waveform $x(n)$ may be characterized by a simple equation, for example,

$$x(n) = \sin\left(\omega_o n\right)$$

Note that $x(n)$ means the value of the signal x at the time index (or space coordinate index) n. In the above example, $x(n)$ is just a sampled sinusoid at a frequency ω_o radians.

For a random signal, second-order statistics are often a sufficient characterization, particularly when we are dealing with linear processing [2]. For example, a stationary white noise signal $w(n)$ is defined by having zero mean,

$$E[w(n)] = 0$$

and an autocorrelation sequence (or autocorrelation function) that is a scaled impulse,

$$R_{ww}(n) \equiv E[w(m)w(m-n)] = \begin{cases} \sigma_w^2, & \text{if } n = 0 \\ 0, & \text{otherwise} \end{cases}$$

where $E[\cdot]$ denotes expected value [3]. Note that σ_w^2 is the variance of each sample of $w(n)$, i.e., $\sigma_w^2 = E[w^2(n)]$.

A white noise signal is a good model for some noises that are encountered in the real world, for example, background noise in radio communications. However, there are other noise sources that may be colored, i.e., their frequency components may have unequal energies. A colored noise $v(n)$ may be generated by passing white noise through a linear system, in the form

$$v(n) = w(n) * h(n) \equiv \sum_{m=-\infty}^{\infty} w(m)h(n-m) \tag{1.1}$$

where $h(n)$ is the impulse response of the system, and $*$ denotes convolution.

Statistically speaking, many signals like speech and images behave like colored noise (at least, in a short time or space scale), so that the model in (1.1) can also be used as a good one for the signals themselves [2]. The autocorrelation function of $x(n)$ as generated in (1.1) is given by [4]

$$R_{xx}(n) \equiv E[x(m)x(m-n)] = \sigma_w^2 \, h(n) * h(-n) \tag{1.2}$$

1.1.2   Power Spectrum

Autocorrelation functions describe the behavior of a random signal in the time domain. We see from (1.2) that if $h(n)$ is a strong filter, i.e., if $h(n)$ is quite different from an impulse, then the autocorrelation of $x(n)$ will also be quite different from an impulse. Following this line of thought, if $h(n)$ is a low-pass filter, with the values of $h(n)$ decaying slowly with n, then $R_{xx}(n)$ will also decay slowly with n, and thus samples of the signal $x(n)$ that are close in time or space will be strongly correlated. Therefore, we conclude that if $h(n)$ is a low-pass filter, then the random signal $x(n)$ will look like a slowly-varying noise. We could reach the same conclusions by looking at the generation of a colored random signal in the frequency domain. For that purpose, we need the definition of the power density spectrum, or simply the power spectrum of a random signal, which is the Fourier transform of its autocorrelation [4],

$$S_{xx}(e^{j\omega}) \equiv \mathcal{F}\{R_{xx}(n)\} = \sum_{n=-\infty}^{\infty} R_{xx}(n) \, e^{-j\omega n}$$

where $\mathcal{F}\{\cdot\}$ is the Fourier transform operator and $j \equiv \sqrt{-1}$. We recall that, $R_{xx}(n)$ being a discrete-time signal, the frequency variable ω is measured in radians and varies from $-\pi$ to π.

The autocorrelation can be recovered simply by the inverse Fourier transform of the power spectrum [4], that is,

$$R_{xx}(n) = \mathcal{F}^{-1}\{S_{xx}(e^{j\omega})\} = \frac{1}{2\pi} \int_{-\pi}^{\pi} S_{xx}(e^{j\omega}) \, e^{j\omega n} \, d\omega$$

We should be careful not to look at the power spectrum as being the same thing as the Fourier transform of the signal itself. Recall that the Fourier transform of $x(n)$ (also referred to as the spectrum of $x(n)$ when the signal is deterministic) is given by [1]

$$X(e^{j\omega}) \equiv \sum_{n=-\infty}^{\infty} x(n) \, e^{-j\omega n} \tag{1.3}$$

and so $X(e^{j\omega})$ is a random variable for every ω because it is a function of the random signal $x(n)$, whereas $S_{xx}(e^{j\omega})$ is a deterministic function of ω. The relationship between these two is simple and intuitive, though. In order to see that, consider a block of M samples of $x(n)$, and assume that (1.3) has been computed over such a block. Then, define the periodogram [2, 4] of that signal block as

$$P_{xx}(e^{j\omega}) \equiv \frac{1}{M}|X(e^{j\omega})|^2 \tag{1.4}$$

It is easy to verify that the power spectrum is simply the expected value of the periodogram

$$S_{xx}(e^{j\omega}) = E[P_{xx}(e^{j\omega})] \tag{1.5}$$

Thus, whereas $X(e^{j\omega})$ is the Fourier spectrum of a particular sample signal of the stochastic process $x(n)$ (and therefore it specifies the frequency components of

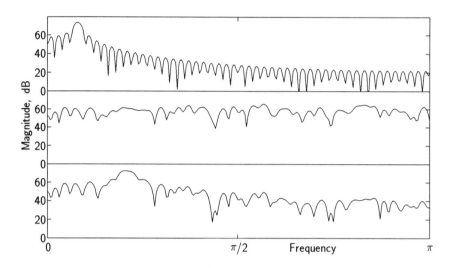

Figure 1.2. Examples of spectra. Top: spectrum of a deterministic sinusoid; middle: spectrum of a white noise; bottom: spectrum of a colored signal.

that particular sample signal), $S_{xx}(e^{j\omega})$ is the expected energy at frequency ω for the whole ensemble of functions in the process $x(n)$. Hence, the power spectrum $S_{xx}(e^{j\omega})$ defines the average frequency distribution of energy for the random signal $x(n)$. These concepts are illustrated in Fig. 1.2.

Looking back at (1.2), its frequency domain equivalent is easy to obtain by means of the convolution theorem [1], with the result

$$S_{xx}(e^{j\omega}) = \sigma_w^2 \, |H(e^{j\omega})|^2 \qquad\qquad (1.6)$$

where $H(e^{j\omega}) = \mathcal{F}\{h(n)\}$. Therefore, for a given signal that we want to model as in (1.1), we can use an average of several periodogram measurements as a good estimate of $S_{xx}(e^{j\omega})$. Finding a filter $h(n)$ that satisfies (1.6) is an approximation problem that can be easily solved with a number of techniques [1, 5, 6].

1.1.3   Autoregressive Models

For a wide class of signals, including speech and images, there is a particular form of
$H(e^{j\omega})$ in (1.6) that is good enough for most applications. This form is an all-pole
filter, with a z-transform that can be written as

$$H(z) = \frac{1}{1 - \sum\limits_{m=1}^{N} a_m \, z^{-m}} \tag{1.7}$$

or, in the Fourier domain ($z = e^{j\omega}$)

$$H(e^{j\omega}) = \frac{1}{1 - \sum\limits_{m=1}^{N} a_m \, e^{-jm\omega}}$$

which corresponds to the following difference equation

$$x(n) = \sum_{m=1}^{N} a_m \, x(n - m) \, + \, \sigma_w^2 \, w(n) \tag{1.8}$$

We note that the above system has no memory of the input, but it memorizes
N previous outputs, which are used together with the current input to generate
the current output. Because of this, the random signal $x(n)$ is referred to as an
autoregressive signal [4] of order N, in short AR(N). A block diagram of the AR(N)
model is shown in Fig. 1.3.

Autoregressive models are quite useful in image and speech processing. Based
on these models, it is possible to compare objectively different signal processing

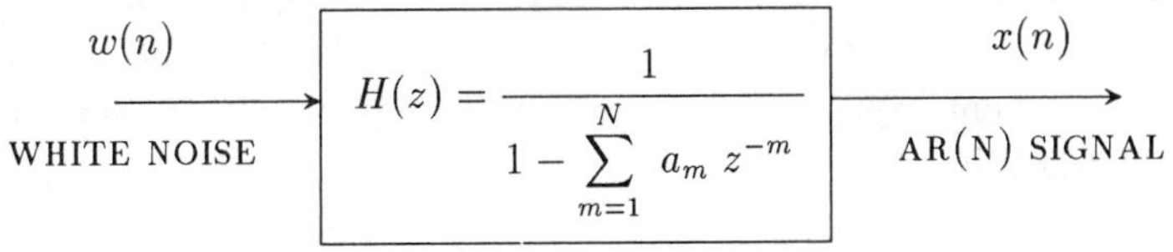

Figure 1.3. Generation of an AR(N) signal

systems. For speech, a model order between $N = 10$ and $N = 20$ is good enough to represent the resonances of the vocal tract [2, 7]. Recently, with the development of the low-delay CELP (code-excited linear prediction) coders for speech, model orders of up to 50 have been suggested [8], with the goal of better modeling the pitch periodicity in female speech.

For images, a first-order model AR(1) is adequate in many cases [2, 9, 10]. It is representative of the image when we look at it as a one-dimensional signal, either by scanning its rows or columns. Strictly speaking, however, an image is a two-dimensional signal of the form $x(n, m)$, where x is represents the light intensity at the point (n, m) of the image plane. So, its autocorrelation function is also two-dimensional, and denoted by $R_{xx}(n, m)$, where n and m now represent the lags in the horizontal and vertical directions. It is usual, in practice, to assume that the image can be modeled as a separable random process [9], i.e., its autocorrelation function can be written in the form

$$R_{xx}(n, m) = R_{xx}(n) \, R_{xx}(m) \tag{1.9}$$

where $R_{xx}(n)$ and $R_{xx}(m)$ are the one-dimensional autocorrelation functions along the rows and columns. The model in (1.9) is valid only approximately for real-world images, but the approximation is good enough for many applications. This separability of the autocorrelation function implies that we can use separable transforms to process the image, without any loss in performance, as we shall discuss in Section 1.2.

1.1.4   Spectral Flatness

After the discussion in the preceding subsections, it is clear that stochastic models for signals, based on autocorrelations and power spectra, should be adequate for many applications. Before we end this section, though, we discuss briefly one important consequence of the random signal models, which is the spectral flatness measure. For a signal $x(n)$ with power spectrum $S_{xx}(e^{j\omega})$, the spectral flatness measure is defined by [2]

$$\gamma_x^2 \equiv \frac{\exp\left(\dfrac{1}{2\pi} \displaystyle\int_{-\pi}^{\pi} \log_e S_{xx}(e^{j\omega}) \, d\omega\right)}{\dfrac{1}{2\pi} \displaystyle\int_{-\pi}^{\pi} S_{xx}(e^{j\omega}) \, d\omega}$$

From the above definition, it follows that [2]

$$0 \leq \gamma_x^2 \leq 1$$

A white noise signal $w(n)$ with $R_{ww}(n) = \sigma_w^2 \, \delta(n)$ has $S_{ww}(e^{j\omega}) = \sigma_w^2$, which leads to $\gamma_w^2 = 1$. A colored signal $x(n)$ will have $\gamma_x^2 < 1$ [2], and this is the reason for the name spectral flatness. The lower the measure, the farther $S_{xx}(e^{j\omega})$ will be from being flat.

Later in the book, we will make use of the concept of the distortion-rate function, which is related to the spectral flatness measure. For signal sources that are stationary and ergodic, the distortion-rate function $D(R)$ specifies the minimum mean-square distortion D that could be achieved at an average rate of R bits per random variable [11]. Therefore, the distortion-rate function provides a lower bound on the fidelity that can be achieved by any coder, for a given communication channel. As intuitively expected, $D(R)$ is a monotonically non-increasing function of R, with $D \rightarrow 0$ as $R \rightarrow \infty$. Thus, it is an invertible function, and its inverse is called the rate-distortion function, $R(D)$.

For a stationary and ergodic source with a Gaussian probability density function, the distortion-rate function is given by [2]

$$D(R) = 2^{-2R} \, \gamma_x^2 \sigma_x^2$$

Ideal (unrealizable) vector quantization of the signal at an average rate of R bits per sample would lead to a distortion D_Q, given by [2]

$$D_Q(R) = 2^{-2R} \, \sigma_x^2 = D(R)/\gamma_x^2 \qquad (1.10)$$

Thus, the spectral flatness measure γ_x^2 is also a measure on how much reduction in mean-square error (over straight signal quantization) we could obtain from the best possible coder. Knowledge of such a bound is certainly quite useful when we are designing a signal coding system because it will tell us how much room there is for improvement in our particular system.

1.2   Block Transforms

Transforms are frequently used in signal processing, and their importance can be verified by the number of books written about the subject, for example [9, 10, 12].

This is because transforms can be employed in many basic signal processing operations, such as filtering, correlation, spectrum estimation, and signal enhancement and coding. In this section we review the definitions and properties of block transforms in general, and then we review the specific properties of the most usual transforms. In Chapter 2 we shall study these transforms in more detail, with a discussion of fast algorithms and applications.

1.2.1   Basic Concepts

Before computing the transform of a given a signal $x(n)$, we must group its samples into blocks. Referring to $\mathbf{x}$ as one of these blocks, we have

$$\mathbf{x} \equiv [x(mM) \;\; x(mM-1) \;\; \ldots \;\; x(mM-M+1)]^T \tag{1.11}$$

where the superscript T denotes transposition (we usually consider signal blocks as column vectors), and m is the block index. The correct notation for $\mathbf{x}$ would be $\mathbf{x}(m)$, to make clear that we have a sequence of signal blocks. However, to avoid confusion with other indices, we will leave the dependence on m implicit. The dimension of $\mathbf{x}$ is M, which is also referred to as the block size or block length.

Consider the following direct linear transformation on $\mathbf{x}$:

$$\mathbf{X} = \mathbf{A}^T \mathbf{x} \tag{1.12}$$

We call $\mathbf{X}$ the transform of $\mathbf{x}$, and $\mathbf{A}$ the transformation matrix, or simply the transform. If the elements in $\mathbf{A}$ are complex, the superscript T denotes conjugate transposition.

A special remark regarding the notation is important here. In many texts, boldface lower case letters are used for vectors and boldface upper case letters for matrices. In dealing with transforms, however, it is also standard practice to use a lower case letter for a signal and the same letter in upper case for its transform. Since the two concepts are not totally compatible, we have decided to follow more closely the second, that is, when we call $\mathbf{x}$ a signal vector, we shall refer to its transform as $\mathbf{X}$.

Recovering the signal block from its transform can be done simply by inverting (1.12):

$$\mathbf{x} = [\mathbf{A}^T]^{-1}\mathbf{X}$$

The equation above is valid for any invertible transform matrix $\mathbf{A}$, but we will consider throughout this book only orthogonal matrices, i.e., those for which

$$\mathbf{A}^T = \mathbf{A}^{-1}$$

Therefore, the inverse transform relation becomes

$$\mathbf{x} = \mathbf{A}\mathbf{X} \tag{1.13}$$

We have used $\mathbf{A}^T$ in the direct transform and $\mathbf{A}$ in the inverse so that the basis vectors (also called basis functions) of the transform are the columns of $\mathbf{A}$. Therefore, we see from (1.12) that the kth element of $\mathbf{X}$ is the inner product of $\mathbf{x}$ with the kth basis function. Furthermore, (1.13) implies that the kth element of $\mathbf{X}$ is also the coefficient of the kth basis function in the linear combination of the basis functions in (1.13) that recovers $\mathbf{x}$ from its transform. For traditional block transforms, the matrix $\mathbf{A}$ is square of order M, so that the number of transform coefficients in $\mathbf{X}$ is the same as the number of signal samples in $\mathbf{x}$.

In practical implementations, orthogonal transforms have many advantages, which come from three basic properties. First, when we choose an orthogonal transform matrix $\mathbf{A}$, both the direct and inverse transform matrices are immediately defined without the need for matrix inversion. Second, the fact the inverse transform operator is just the transpose of the direct operator means that, given a flowgraph for the direct transform, the inverse transform can be implemented by transposing the flowgraph, i.e., by running it backwards [10, 13]. Finally, orthogonal transforms conserve energy, that is,

$$\|\mathbf{X}\| = \|\mathbf{x}\|$$

where $\|\mathbf{x}\|$ is the Euclidean norm of $\mathbf{x}$, that is,

$$\|\mathbf{x}\| \equiv \sum_{k=1}^{M} |[\mathbf{x}]_k|^2$$

where $[\mathbf{x}]_k$ denotes the k-th element of the vector $\mathbf{x}$.

Two other properties of orthogonal transforms that we use later in the book are that all eigenvalues of $\mathbf{A}$ lie on the unit circle of the complex plane [14], and that there are $M(M-1)/2$ degrees of freedom in choosing the M^2 elements of $\mathbf{A}$. This last property can be easily obtained by recalling that the orthogonality of $\mathbf{A}$ implies

$$\mathbf{A}^T\mathbf{A} = \mathbf{A}\mathbf{A}^T = \mathbf{I} \tag{1.14}$$

where $\mathbf{I}$ is the identity matrix. Since there are $M(M+1)/2$ distinct equations in (1.14) on the M^2 elements of $\mathbf{A}$, we have $M(M-1)/2$ degrees of freedom.

Each choice of $\mathbf{A}$ leads to a different transform. Of course, the number of possible choices is infinite, but some of them lead to matrices with special properties that are useful in signal processing applications. In the following subsections we discuss briefly the most commonly used transforms, and the fundamental properties of each.

1.2.2   Discrete Fourier Transform

The discrete Fourier transform (DFT) [1, 5, 13] is defined by basis functions that are complex sinusoids with frequencies varying linearly from 0 to π, in the form

$$a_{nk} = \sqrt{\frac{1}{M}} \, \exp\left(j\frac{2\pi kn}{M}\right) \tag{1.15}$$

where a_{nk} means the element of $\mathbf{A}$ in the nth row and kth column or, equivalently, the nth sample of the kth basis function. As is usual in most of the literature dealing with transforms, we assume that the indices n and k vary from 0 to $M-1$. The scaling factor $\sqrt{1/M}$ is chosen to keep the orthogonality of $\mathbf{A}$. We note that the transform coefficient index k can be viewed as a frequency index.

The definition of the DFT basis functions in (1.15) has two basic interpretations [13]. In the first, the DFT corresponds to a discretization of the Fourier transform in (1.3), in the sense that the transform coefficients of the DFT are samples of the Fourier transform, in the form [1, 5]

$$[\mathbf{X}]_k = X(e^{j\omega})\big|_{\omega=2\pi k/M} \tag{1.16}$$

where $[\mathbf{X}]_k$ is the kth element of $\mathbf{X}$, i.e., the kth transform coefficient. In the second interpretation, the DFT coefficients in $\mathbf{X}$ can be viewed as the discrete Fourier series of a bandlimited continuous waveform $x(t)$, which has been periodically sampled to obtain the signal vector $\mathbf{x}$.

Whatever interpretation we feel to be more appropriate, it is clear from (1.13) and (1.15) that the DFT represents the signal vector $\mathbf{x}$ as a linear combination of harmonically-related complex sinusoids.

Assuming that the input block $\mathbf{x}$ is real, (1.15) leads to the well-known property

$$[\mathbf{X}]_{M-k} = [\mathbf{X}]_k^* \implies |[\mathbf{X}]_{M-k}| = |[\mathbf{X}]_k| \qquad (1.17)$$

where the superscript $*$ denotes complex conjugation. For M even, we see that $[\mathbf{X}]_0$ and $[\mathbf{X}]_{M/2}$ are purely real (zero imaginary part). It is important to note that the relationships in (1.17) are a consequence of the fact that there are only $M/2$ distinct frequencies in the DFT basis functions defined in (1.15). The kth and $(M-k)$th basis functions have the same frequency and their phases differ by $\pi/2$.

One of the main uses for the DFT in signal processing is in spectrum estimation [1, 5, 13] because the DFT of an M-point signal block can be viewed as an approximation to the Fourier transform of the signal at equally spaced frequencies, according to (1.16). By increasing the block length M, we can improve the frequency resolution of the DFT. If the signal $x(n)$ is random, the DFT of each block will also be random, but an average of several periodograms [defined in (1.4)] will be a good estimate of the power spectrum $S_{xx}(e^{j\omega})$ [4]. There are many other techniques for spectrum estimation [15, 16], but they are beyond the scope of this book. The main disadvantage of using the DFT for power spectrum estimation is that with a block of length M we will have information about only $M/2$ frequencies. There are other sinusoidal transforms that do not have this disadvantage, as we shall see in the following.

Another important use for the DFT in signal processing is in filtering, either fixed or adaptive [17]. The reason for that is the convolution theorem [1, 5], which states that if $y(n) = x(n) * h(n)$, then

$$Y(e^{j\omega}) = X(e^{j\omega})H(e^{j\omega})$$

This property is not immediately shared by the DFT; that is, if $y(n) = x(n) * h(n)$ and $\mathbf{y}$, $\mathbf{x}$, and $\mathbf{h}$ are blocks of these signals, we have, in general,

$$\mathbf{Y} \neq \mathbf{X} \times \mathbf{H} \qquad (1.18)$$

where $\times$ denotes the Hadamard product operator [18], i.e., the product performed element by element. If the blocks $\mathbf{x}$ and $\mathbf{h}$ are padded with enough zeros, though, then (1.18) becomes an equality. Thus, the convolution can be performed block by block with a direct DFT, a Hadamard product, and an inverse DFT, in an overlap-add or overlap-save procedure [1], as we see in the next chapter.

The use of the DFT in spectrum estimation and signal filtering is further justified by the existence of fast algorithms for the DFT, collectively known as the fast Fourier transform (FFT) [10, 13]. We return to these issues in Chapter 2.

1.2.3   Discrete Hartley Transform

The discrete Hartley transform (DHT) [19] is a real version of the DFT. Its basis functions are obtained simply by adding the real and imaginary parts of the DFT, in the form

$$a_{nk} = \sqrt{\frac{1}{M}} \operatorname{cas}\left(\frac{2\pi nk}{M}\right) \tag{1.19}$$

where $\operatorname{cas}(\theta) \equiv \cos(\theta) + \sin(\theta)$. Note that, as was the case with the DFT, there are $M/2$ distinct sinusoids in (1.19). The kth and $(M - k)$th basis functions have the same frequency, but are 90 degrees out of phase. Some of the basis functions of the DHT are shown in Fig. 1.4, for $M = 32$. The discrete Hartley transform was also independently derived in [20], where it was called type-I discrete W transform.

There are many more similarities than differences between the DHT and the DFT, and anything that can be done with the DFT can also be done with the DHT [19], including spectrum estimation and convolution. In fact, given the DHT coefficients of a signal block, it is easy to compute the corresponding DFT coefficients, and vice versa [19]. The DHT has two advantages, though: first, it is a real transform, so no complex arithmetic is necessary; second, we see from (1.19) that the DHT matrix is symmetric, and therefore the direct and inverse transforms are identical. This second point is quite important because it means that in signal filtering and coding applications, which require both the direct and inverse transforms, only one routine for computing the DHT is required, since it will also perform the inverse DHT.

Fast algorithms for the DHT, collectively referred to as FHT algorithms, also exist, and they have approximately the same computational complexity of the corresponding FFT algorithms, as we see in Chapter 2.

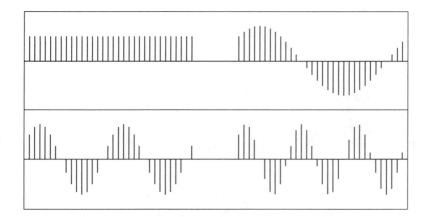

Figure 1.4. The first four basis functions of the DHT, for $M = 32$.

1.2.4   Karhunen-Loève Transform

The Karhunen-Loève transform (KLT) [2], which is also referred to as the Hotelling transform [21], is an optimal transform from an statistical viewpoint. This is because the KLT is the unique orthogonal transform that will produce a set of uncorrelated coefficients from a nonwhite signal. We will see later why this decorrelation is important. For now, let us consider the covariance matrix $\mathbf{R_{xx}}$ of the input signal block $\mathbf{x}$

$$\mathbf{R_{xx}} \equiv E[\mathbf{xx}^T] \tag{1.20}$$

Note that in the above definition we have already considered that $\mathbf{x}$ is composed of samples of a zero-mean signal. From (1.11) and (1.20), it is clear that the elements of the covariance matrix are samples of the autocorrelation function of $x(n)$, that is,

$$[\mathbf{R_{xx}}]_{nk} = R_{xx}(n - k) = R_{xx}(k - n)$$

The equation above shows that $\mathbf{R_{xx}}$ is a symmetric and Toeplitz matrix (all diagonals have identical entries). Furthermore, it is well known that covariance matrices are positive semidefinite [2, 21], i.e., all of its eigenvalues are real and nonnegative.

Given the input block $\mathbf{x}$ with covariance $\mathbf{R_{xx}}$, the covariance of the transformed block $\mathbf{X}$ is easily obtained from (1.12) as

$$\mathbf{R_{XX}} = \mathbf{A}^T \, \mathbf{R_{xx}} \, \mathbf{A}$$

The KLT transform is defined by the $\mathbf{A_{KL}}$ matrix that will diagonalize $\mathbf{R_{XX}}$, in the form

$$\mathbf{R_{XX}} = \mathbf{A}_{KL}^T \, \mathbf{R_{xx}} \, \mathbf{A_{KL}} = \text{diag} \, \{\lambda_0, \lambda_1, \ldots, \lambda_{M-1}\} \qquad (1.21)$$

It is well known from matrix theory [14] that the matrix $\mathbf{A_{KL}}$ which will satisfy the diagonalization condition in (1.21) is the orthogonal matrix whose columns are the eigenvectors of $\mathbf{R_{xx}}$, with the diagonal entries of $\mathbf{R_{XX}}$ being the eigenvalues, that is,

$$\mathbf{R_{xx}} \, [\mathbf{A}]_{.k} = \lambda_k \, [\mathbf{A}]_{.k}$$

where $[\mathbf{A}]_{.k}$ is the kth column of $\mathbf{A}$. Thus, the basis functions of the KLT are the eigenvectors of the covariance matrix of the input signal. For AR(1) signals, such eigenvectors have the property of being sinusoids [2], with frequencies that are not equally distributed in the unit circle. The phases of these sinusoids are determined from the fact that eigenvectors of Toeplitz matrices have either even or odd symmetry [22]. Some of the basis functions of the KLT are shown, for $M = 32$, in Fig. 1.5, for an AR(1) signal, and in Fig. 1.6, for an AR(12) model derived from a female speech signal. It is interesting to note that the basis functions shown in Fig. 1.6 are quite similar to the KLT functions obtained in [23] for different speech signals, which is a confirmation that statistical modeling can be helpful in getting the average properties of most families of signals.

Note that (1.21) implies that the transforms coefficients (the elements of $\mathbf{X}$) are uncorrelated, and that their variances are given by

$$\text{Var} \, \{[\mathbf{X}]_k\} \equiv \sigma_k^2 = \lambda_k$$

By diagonalizing the covariance matrix of the transform coefficients, the KLT also maximizes the energy compaction in $\mathbf{X}$, in the sense that the above variance distribution of $\mathbf{X}$ is the most concentrated (there will be a few transform coefficients with large variances, and most coefficients will have small variances). This implies that the product of all variances is minimized [2].

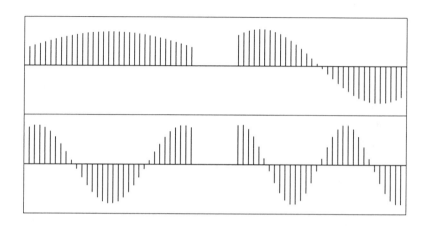

Figure 1.5. The first four basis functions of the KLT, $M = 32$, for an AR(1) signal with intersample correlation coefficient of 0.9.

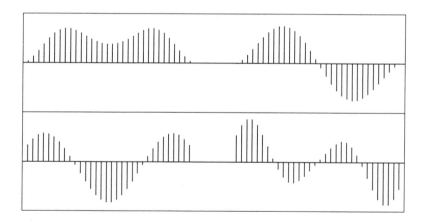

Figure 1.6. The first four basis functions of the KLT, $M = 32$, for an AR(12) model derived from a female speech signal.

This energy concentration has an important implication for signal coding applications. In transform coding (TC) [9], instead of quantizing the samples of the signal with a desired number of bits per sample (which is referred to as pulse-code modulation, or PCM), we can perform the quantization on the transform coefficients, with a different number of bits for each. The total number of bits should be the same in both cases. It is well known [2, 9] that a lower mean-square error will result from quantizing the transform coefficients. Assuming scalar quantizers, the reduction in TC mean-square error over PCM is given, as we see in Chapter 2, by [2, 9]

$$G_{\text{TC}} = \frac{\dfrac{1}{M} \sum_{k=0}^{M-1} \sigma_k^2}{\left(\displaystyle\prod_{k=0}^{M-1} \sigma_k^2 \right)^{1/M}}$$

which is referred to as the transform coding gain. Note that the numerator in the equation above is in fact the average energy in the transformed block $\mathbf{X}$, which is identical to the input signal variance, due to the total energy conservation of any orthogonal transform. The denominator, however, is the geometric mean of the transform coefficient variances, which is minimized by the KLT. Therefore, the KLT maximizes G_{TC}, and so it is the optimal transform for TC.

With the KLT, transform coding is an asymptotically optimal coding system because [2]

$$\lim_{M \to \infty} G_{\text{TC}} = 1/\gamma_x^2$$

Thus, the reduction in mean-square error for TC reaches the maximum value of $1/\gamma_x^2$ predicted by the distortion-rate function in (1.10).

From the discussion above it is clear that the KLT should ideally be the transform of choice in a signal coding system, due to its optimality. However, the KLT is in fact seldom used in practice for two reasons. First, the KLT is signal-dependent, so we need to have a good model for the signal statistics, which is not always available. Furthermore, the signal statistics might change with time, and real-time computation of covariance matrices and their eigenvectors is virtually impossible for most applications. Second, the KLT matrix $\mathbf{A}_{\text{KL}}$ has no particular structure, so the transform operator in (1.12) would require M^2 multiplications and additions (compared to the $O(M \log M)$ operations required for the FFT or FHT, as discussed in Chapter 2).

An important asymptotic property of the KLT is that for an AR(1) signal, as the correlation coefficient between adjacent samples $R_{xx}(1)/R_{xx}(0)$ tends to one, the basis functions of the KLT become sinusoids at frequencies equally-spaced in the interval $[0, \pi)$ [2, 9]. These sinusoids assume precisely the form of the discrete cosine transform, which we discuss next.

1.2.5 Discrete Cosine Transform

The discrete cosine transform (DCT) [10] was originally derived from Chebyshev polynomials [24], but with the aim of approximating the eigenvectors of the autocorrelation matrix of an AR(1) signal block. The asymptotic equivalence of the DCT and the KLT was later demonstrated [25]. The definition of the DCT basis functions is

$$a_{nk} = c(k) \sqrt{\frac{2}{M}} \cos\left[\left(n + \frac{1}{2}\right) \frac{k\pi}{M}\right] \tag{1.22}$$

where

$$c(k) \equiv \begin{cases} 1/\sqrt{2} & \text{if } k = 0 \\ 1 & \text{otherwise} \end{cases}$$

A plot of the first four basis functions of the DCT is shown in Fig. 1.7, for $M = 32$. We should point out that there are other DCT definitions [10], but the DCT corresponding to (1.22) is the most employed in practice (for example, IEEE standard no. 1180 [26] is based on that definition).

The main difference between the DCT and the DFT or DHT is that, when the frequency index k varies from 0 to $M - 1$, there are M different frequencies in the interval $[0, \pi]$. Therefore, a spectral analysis performed with the DCT will produce twice the number of frequency bands as does the DFT or DHT, but with the disadvantage that the DCT spectrum has no magnitude/phase interpretation. Furthermore, if the input signal has a strong component at a particular DCT frequency, but that component is 90 degrees out of phase with respect to the DCT basis function, then the corresponding DCT coefficient will be zero. Thus, the DCT of a single block may not be much useful for spectral analysis, but a DCT-based

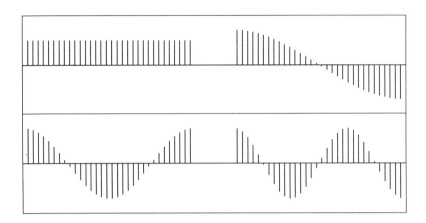

Figure 1.7. The first four basis functions of the DCT, for $M = 32$.

periodogram would also be a good estimate of the Fourier spectrum, as we see in Chapter 2.

The DCT can be used for convolution and correlation, because it satisfies a modified shift property [10], but with a higher computational complexity than DFT-based convolution. Therefore, the major uses of the DCT are in transform coding and frequency-domain adaptive filtering. The use of the DCT in signal coding is so widespread that the DCT is a candidate for becoming a CCITT standard for videophone and other image coding applications [10]. We will discuss the applications and fast algorithms of the DCT in Chapter 2.

1.2.6   Type-IV Discrete Cosine Transform

The DCT-IV was introduced in [20] as an alternative transform for spectral analysis. It is obtained by shifting the frequencies of the DCT basis functions in (1.22) by $\pi/2M$, in the form

$$a_{nk} = \sqrt{\frac{2}{M}} \, \cos\left[\left(n + \frac{1}{2}\right)\left(k + \frac{1}{2}\right)\frac{\pi}{M}\right] \tag{1.23}$$

Note that, unlike the DCT, the scaling factor is identical for all basis functions. The first four basis functions of the DCT-IV are shown in Fig. 1.8, for $M = 32$.

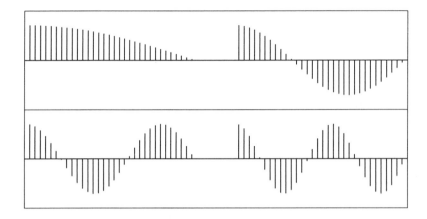

Figure 1.8. The first four basis functions of the DCT-IV, for $M = 32$.

Because of the frequency shift of the DCT-IV basis functions, when compared to the DCT basis, the DCT-IV is not as useful as the DCT for signal coding, as we will see in the next chapter. However, the DCT-IV can be successfully applied to spectrum estimation and adaptive filtering. Furthermore, the DCT-IV can be used as a building block for fast algorithms for the DCT and for lapped transforms.

Related to the DCT-IV is the DST-IV, defined by [20]

$$a_{nk} = \sqrt{\frac{2}{M}} \, \sin\left[\left(n + \frac{1}{2}\right)\left(k + \frac{1}{2}\right)\frac{\pi}{M}\right] \qquad (1.24)$$

Note that if we take both the DCT-IV and the DST-IV of an input signal, then we would have M frequencies of spectrum information in the interval $[0, \pi]$, which is twice the resolution of the DFT or DHT. Magnitude and phase information would then be preserved because we have projections on cosines with the DCT-IV and on sines with the DST-IV.

1.2.7 Other Transforms

Except for the KLT for non-AR(1) signals, all of the previously considered transforms have basis functions that are sampled sinusoids. In applications where the computational complexity of sinusoidal transforms may be prohibitive, other transforms have been considered, such as the Walsh-Hadamard, Haar, slant, and high-correlation transform [9, 21, 27]. All of these transforms have fast algorithms, which are less complex than those of the sinusoidal transforms. However, the easier computation is achieved at the cost of worse performance in signal filtering or coding.

Thus, with the availability of DSP hardware of ever increasing computing speed and decreasing cost, the nonsinusoidal transforms are now employed only in rare situations. Therefore, we shall not discuss them further in this book.

1.2.8   Two-Dimensional Transforms

Our discussion of block transforms throughout this section has assumed that the input signal block is one-dimensional. In image coding applications, however, the blocks are actually $M \times M$ matrices, and the transform operator is a four-dimensional tensor. In these applications [9, 21, 27], there is no advantage in using general orthogonal tensors for two-dimensional (2-D) transforms, so separable transforms can always be employed. Then, the transform tensor for 2-D signals $\mathbf{B}$ can be put in the separable form

$$[\mathbf{B}]_{lmrs} = [\mathbf{A}]_{lr} \times [\mathbf{A}]_{ms}$$

where $\times$ denotes the Kronecker product [14].

The equation above states that the 2-D transform operator can be applied in two steps: transform all the rows in the block with a 1-D transform, and then transform all columns in the transformed block with the same 1-D transform (we could start with transforming the columns, and then the rows). Therefore, all the discussion on block transforms that we have carried on until now, as well as the subsequent discussions on lapped transforms, remain valid for 2-D transforms. We simply use separable transforms.

1.3   Development of Lapped Transforms

The basic motivation behind the development of lapped transforms comes from one of the major disadvantages of traditional block transforms: the *blocking effects*, which are discontinuities in the reconstructed signal. In transform coding or block filtering, we start by removing a block of M samples from the input signal. The block is then transformed, using one of the transforms discussed in the previous section. The transform coefficients are then processed according to the application: in transform coding [2, 9], they are quantized, either with scalar or vector quantization [28], and in fixed or adaptive filtering they are multiplied by appropriate scaling factors [29]. Finally, the inverse transform is computed, and the reconstructed signal block is appended to the output.

Because of the independent processing of each block, some of the coding errors will produce discontinuities in the signal because the final samples of one block will, most likely, not match with the first samples of the next and thus generate the blocking effects. In image coding, the blocking effects produce a reconstructed image that seems to be built of small tiles [10, 28], whereas in speech coding the reconstructed speech will have an audible periodic noise [30].

Several approaches towards the reduction of blocking effects were developed. Among them, prefiltering and postfiltering [31, 32] and the short-space Fourier transform (SSFT) [33]. The filtering techniques have the disadvantage of reducing the transform coding gain, and producing a visible low-pass effect around the block boundaries, but they have as an advantage a computational overhead of less than 10 percent. With the SSFT, there are no blocking effects, because the SSFT basis functions are theoretically infinite in length. The disadvantage with the SSFT is the appearance of ringing around the edges of an image. Also, the SSFT can only be efficiently applied to finite-length signals, such as images [33], or to periodic signals.

The first lapped transform that was designed with the aim of reducing the blocking effects was the lapped orthogonal transform (LOT) [34, 35]. The key to the development of the LOT was the recognition that the blocking effects are really caused by the discontinuities on the basis functions of the transforms. These discontinuities do not exist when we are only looking at a single block, but they do appear when we realize that each block is reconstructed as a weighted combination of the basis functions. When each of these functions is pasted in its position in the output stream of samples, a discontinuity is caused. To better visualize this result, imagine that only a single reconstructed block is sent to the output. Then, the signal will jump from and to zero before and after the block. Another motivation in the same direction can be obtained by looking at Fig. 1.6, where we see that the first basis functions of the KLT of a physical signal do not have sharp discontinuities at their boundaries.

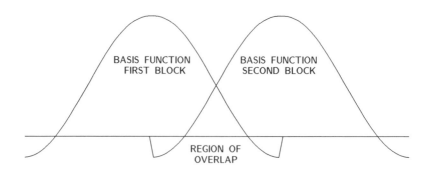

Figure 1.9. The overlapping of the basis functions in the LOT.

After better understanding of the generation of blocking effects, the development of the LOT started with the fundamental question: can the basis functions of the transform be made longer than the transform length? In this way, the basis functions could have a smoother transition to and from zero at their ends. Thus, if the number of transform coefficients is M, the basis functions would have more than M samples. Nevertheless, each new input block would still be taken every M samples, and so the basis functions of the LOT would be projected not only on the current block, but also on the neighboring blocks, on both sides. Therefore, the basis functions from one block and one of its neighboring blocks would overlap, and this is the reason for the name *lapped*. This property is illustrated in Fig. 1.9.

If we want to use the same basis functions for analysis (direct transform) and synthesis (inverse transform), the overlapping portions of the basis functions of neighboring blocks must be orthogonal. This is the reason for the LOT denomination. Fortunately, the number of constraints incurred by these extra orthogonality restrictions is less than the number of extra degrees of freedom added by increasing the lengths of the basis functions. Therefore, LOT basis functions can be constructed.

Initially, the LOT basis functions were constrained to have their lengths equal to twice the block size, i.e., $2M$. Then, the LOT of a signal block $\mathbf{x}$ can be obtained by [34, 35]

$$\mathbf{X} = \mathbf{P}^T \mathbf{x} \qquad (1.25)$$

where $\mathbf{P}$ is the lapped transform matrix, with the LT basis functions as its columns, and $\mathbf{x}$ is an extended signal block having $2M$ samples

$$\mathbf{x} = [x(mM - 2M + 1) \ \ x(mM - 2M + 2) \ \ \ldots \ \ x(mM - 1) \ \ x(mM)]^T (1.26)$$

with m being the block index. Thus, it is clear that each sample of the input signal $x(n)$ will be used in two blocks. Note that we have made implicit again the dependence of $\mathbf{x}$ on m to simplify the notation.

In order to allow the reconstruction of the original signal $x(n)$ from the sequence of LOT blocks $\mathbf{X}$, the LOT matrix $\mathbf{P}$ should satisfy the following orthogonality conditions [34, 36]

$$\mathbf{P}^T\mathbf{P} = \mathbf{I} \quad \text{and} \quad \mathbf{P}^T\mathbf{W}\mathbf{P} = 0 \qquad (1.27)$$

where $\mathbf{I}$ is the identity matrix and $\mathbf{W}$ is the M-sample shift operator defined by

$$\mathbf{W} \equiv \begin{pmatrix} \mathbf{0} & \mathbf{I} \\ \mathbf{0} & \mathbf{0} \end{pmatrix} \qquad (1.28)$$

where the identity matrix above is of order M.

We note that the LOT definition in (1.27) is quite general. For example, a block transform $\mathbf{A}$ can also be viewed as a LOT, with $\mathbf{P}^T = [\mathbf{0} \ \mathbf{A}^T \ \mathbf{0}]$. In fact, when we look at a block transform in this way, the discontinuities in the basis functions become clear. We want to find a matrix $\mathbf{P}$ that has nonzero entries and is a valid LOT. The first approach toward that goal [34] was to solve the problem in (1.27) in an iterative way, starting with the first basis function that led to the maximum σ_0^2, and then solving a nonlinear optimization problem to find the second basis function that maximized σ_1^2, under the constraints in (1.27), and so on for the third and remaining basis functions. In this way, the LOT with the maximum G_{TC} would be obtained. This approach did not lead to good results, though, because the procedure may converge to local optima.

Another approach to the derivation of a LOT is to start within a subspace of valid LOTs [36, 37], and solve an eigenvector problem to derive an optimal LOT. In this way, the equivalent of a Karhunen-Loève LOT is obtained. An example of an optimal LOT derived in this way is shown in Fig. 1.10.

The first experiments with LOT coding of images [35, 38] were quite encouraging. By simply replacing the DCT by the LOT in a transform coder, the blocking

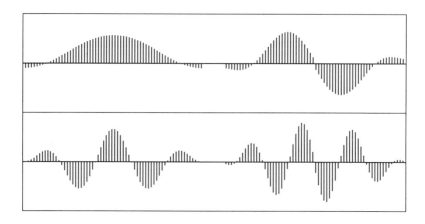

Figure 1.10. The first four basis functions of an optimal LOT, for $M = 32$.

effects were virtually eliminated. However, the optimal LOTs obtained with the aforementioned methods had the disadvantage of not being fast computable. After the first suggestion of a fast LOT algorithm in [37, 38], the LOT became a serious candidate for replacing the DCT in transform coding. Further experiments with the LOT also have demonstrated its superior performance in adaptive speech coding [30, 39], and data recovery in packet video with packet loss [40].

Later, it became clear that the family of lapped transforms could be expanded by removing two of the restrictions on their basis functions. First, the even or odd symmetry of the functions, initially assumed in the developments in [34, 37] was relaxed, leading to the development of the modulated lapped transform (MLT) [39]. Second, the length of the basis functions was not restricted to $2M$, so that lapped transforms with arbitrarily long basis functions could be designed, as first demonstrated in [41].

The purpose of this section was to give a brief review of the history of the development of lapped transforms. Throughout the book, all of the details of the theoretical properties and fast algorithms for LTs will be presented, as well as many examples of practical applications of LTs.

References

[1] Oppenheim, A. V., and R. Schafer, *Discrete-Time Signal Processing*, Englewood Cliffs, NJ: Prentice-Hall, 1989.

[2] Jayant, N. S., and P. Noll, *Digital Coding of Waveforms*, Englewood Cliffs, NJ: Prentice-Hall, 1984.

[3] Drake, A. W., *Fundamentals of Applied Probability Theory*, New York: McGraw-Hill, 1967.

[4] Papoulis, A., *Probability, Random Variables, and Stochastic Processes*, New York: McGraw-Hill, 2nd ed., 1984.

[5] De Fatta, D., J. G. Lucas, and W. S. Hodgkiss, *Digital Signal Processing: A System Design Approach*, New York: Wiley, 1988.

[6] IEEE DSP Committee, ed., *Programs for Digital Signal Processing*, New York: IEEE Press, 1979.

[7] Rabiner, L. R., and R. W. Schafer, *Digital Processing of Speech Signals*, Englewood Cliffs, NJ: Prentice-Hall, 1978.

[8] Chen, J. H., "High-quality 16 kb/s speech coding with a one-way delay less than 2 ms," *IEEE Intl. Conf. Acoust., Speech, Signal Processing*, Albuquerque, NM, pp. 453–457, Apr. 1990.

[9] Clarke, R. J., *Transform Coding of Images*, London: Academic Press, 1985.

[10] Rao, K. R., and P. Yip, *Discrete Cosine Transform: Algorithms, Advantages, Applications*, New York: Academic Press, 1990.

[11] Berger, T., *Rate Distortion Theory*, Englewood Cliffs, NJ: Prentice-Hall, 1971.

[12] Beauchamp, K. G., *Transforms for Engineers: A Guide to Signal Processing*, New York: Oxford University Press, 1987.

[13] Brigham, E. O., *The Fast Fourier Transform*, Englewood Cliffs, NJ: Prentice-Hall, 1974.

[14] Gantmacher, F. R., *The Theory of Matrices, Vol. I*, New York: Chelsea, 1977.

[15] Childers, D. G., ed., *Modern Spectrum Analysis*, New York: IEEE Press, 1978.

[16] Gardner, W. A., *Statistical Spectral Analysis: A Nonprobabilistic Approach*, Englewood Cliffs, NJ: Prentice-Hall, 1988.

[17] Widrow, B., and S. D. Stearns, *Adaptive Signal Processing*, Englewood Cliffs, NJ: Prentice-Hall, 1985.

[18] Graham, A., *Kronecker Products and Matrix Calculus: with Applications*, Chichester, England: Ellis Horwood, 1981.

[19] Bracewell, R. N., "Discrete Hartley transform," *J. Opt. Soc. Am.*, vol. 73, Dec. 1983, pp. 1832–1835.

[20] Wang, Z., "Fast algorithms for the discrete W transform and for the discrete Fourier transform," *IEEE Trans. Acoust., Speech, Signal Processing*, vol. ASSP-32, Aug. 1984, pp. 803–816.

[21] Gonzalez, R. C., and P. Wintz, *Digital Image Processing*, Reading, MA: Addison-Wesley, 1977.

[22] Makhoul, J., "On the eigenvectors of symmetric Toeplitz matrices," *IEEE Trans. Acoust., Speech, Signal Processing*, vol. ASSP-29, Aug. 1981, pp. 868–872.

[23] Campanella, S. J., and G. S. Robinson, "A comparison of orthogonal transforms for digital speech processing," *IEEE Trans. Commun. Tech.*, vol. COM-19, Dec. 1971, pp. 1045–1050.

[24] Ahmed, N., T. Natarajan, and K. R. Rao, "Discrete cosine transform," *IEEE Trans. Comput.*, vol. C-23, Jan. 1974, pp. 90–93.

[25] Clarke, R. J., "Relation between the Karhunen-Loève and cosine transforms," *Proc. IEE,* Pt. F, vol. 128, Nov. 1981, pp. 359–360.

[26] IEEE Circuits and Systems Society, "IEEE standard no. 1180 – specifications for the implementation of 8×8 inverse discrete cosine transform," Tech. Rep., IEEE, New York, March 1991.

[27] Pratt, W. K., ed., *Digital Image Processing*, New York: Wiley, 1978.

[28] Netravali, A. N., and B. G. Haskell, *Digital Pictures: Representation and Compression*, New York: Plenum, 1989.

[29] Cowan, C. F. N., and P. Grant, eds., *Adaptive Filters*, Englewood Cliffs, NJ: Prentice-Hall, 1985.

[30] Malvar, H. S., and R. Duarte, "Transform/subband coding of speech with the lapped orthogonal transform," *IEEE Intl. Symp. Circuits Syst.*, Portland, OR, pp. 1268–1271, May 1989.

[31] Reeve, H. C., and J. S. Lim, "Reduction of blocking effect in image coding," *IEEE Intl. Conf. Acoust., Speech, Signal Processing*, Boston, MA, pp. 1212–1215, March 1983.

[32] Malvar, H. S., "A pre- and post-filtering technique for the reduction of blocking effects," *Picture Coding Symp.*, Stockholm, Sweden, June 1987.

[33] Hinman, B. L., J. G. Bernstein, and D. H. Staelin, "Short-space Fourier transform image processing," *IEEE Intl. Conf. Acoust., Speech, Signal Processing*, San Diego, CA, pp. 4.8.1–4.8.4, March 1984.

[34] Cassereau, P., *A New Class of Optimal Unitary Transforms for Image Processing*, Master's thesis, Mass. Inst. Tech., Cambridge, MA, May 1985.

[35] Cassereau, P. M., D. H. Staelin, and G. de Jager, "Encoding of images based on a lapped orthogonal transform," *IEEE Trans. Commun.*, vol. 37, Feb. 1989, pp. 189–193.

[36] Malvar, H. S., *Optimal Pre- and Post-filters in Noisy Sampled-Data Systems*, PhD thesis, Mass. Inst. Tech., Cambridge, MA, Sep. 1986.

[37] Malvar, H. S., and D. H. Staelin, "The LOT: transform coding without blocking effects," *IEEE Trans. Acoust., Speech, Signal Processing*, vol. 37, Apr. 1989, pp. 553–559.

[38] Malvar, H. S., and D. H. Staelin, "Reduction of blocking effects in image coding with a lapped orthogonal transform," *IEEE Intl. Conf. Acoust., Speech, Signal Processing*, New York, pp. 781–784, Apr. 1988.

[39] Malvar, H. S., "Lapped transforms for efficient transform/subband coding," *IEEE Trans. Acoust., Speech, Signal Processing*, vol. 38, June 1990, pp. 969–978.

[40] Haskell, P., and D. Messerschmitt, "Reconstructing lost video data in a lapped orthogonal transform based coder," *IEEE Intl. Conf. Acoust., Speech, Signal Processing*, Albuquerque, NM, pp. 1985–1988, Apr. 1990.

[41] Malvar, H. S., "Modulated QMF filter banks with perfect reconstruction," *Electron. Lett.*, vol. 26, no. 13, June 1990, pp. 906–907.

Chapter 2

Applications of Block Transforms

In this chapter we discuss in more detail the classical block transforms that were presented in Chapter 1. We start by looking at the most common applications of block transforms in signal processing: filtering, spectrum estimation, and coding, in Sections 2.1 to 2.3. Other applications are briefly discussed in Section 2.4. We certainly cannot cover all of those applications in great detail (there are entire books written on each one of them). Our purpose is to make clear how orthogonal transforms can be applied. This is essential to enable us to judge whether a lapped transform can be employed instead of a block transform for a given application.

In most cases, the time consumed in performing the transforms is an important issue. Simply computing the transforms by the matrix multiplications defined in Chapter 1 is unacceptable in virtually any application because it would generally result in an excessive requirement on the computational power of the processing system. Thus, in most applications the transforms are computed by fast algorithms. We believe that it is important for the system design engineer to have fast procedures for block transforms readily available, so the fastest known algorithms for the computation of the most commonly used block transforms are discussed in Section 2.5. Computer programs implementing these algorithms can be found in Appendix B.

2.1   Signal Filtering

The basic operation in digital signal processing is filtering. Assuming that $x(n)$ is
the input and $y(n)$ the output of a linear shift-invariant filter with impulse response
$h(n)$, it is well known [1] that the relationship between $y(n)$ and $x(n)$ is given by

$$y(n) = x(n) * h(n) \equiv \sum_{m=-\infty}^{\infty} h(m)x(n-m)$$

where $*$ is the convolution operator. In many practical cases, a finite impulse
response filter (FIR) can be used. For a FIR filter, there are indices m_1 and m_2
such that $h(m) = 0$ if $m \notin [m_1, m_2]$. Without loss of generality, we can assume
that $m_1 = 0$ and $m_2 = L - 1$, where L is the filter length, so that the convolution
becomes the finite sum

$$y(n) = \sum_{m=0}^{L-1} h(m)x(n-m) \tag{2.1}$$

The above equation can also be viewed as the inner product

$$y(n) = \mathbf{h}^T \mathbf{x}_n \tag{2.2}$$

where the vector $\mathbf{h}$ contains the filter coefficients, and the vector $\mathbf{x}_n$ is a signal block
containing L samples, from $x(n - L + 1)$ to $x(n)$.

Viewing a FIR filtering operation as an inner product is quite useful when we are
working with block signal processing. In fact, comparing equations (2.2) with (1.12),
we see that a transform operation is equivalent to the computation of the outputs
of many filters on the same input signal block $\mathbf{x}$. Each filter impulse response is a
column of the transform matrix $\mathbf{A}$. Therefore, a block transform can also be viewed
as a filter bank, and this connection will be further discussed in the next chapter.

2.1.1   Efficient FIR Filtering

The simplest practical implementation of the FIR filter in (2.1) and (2.2) is the direct
form [1]. In this approach, whenever an input sample $x(n)$ becomes available, it is
shifted into block $\mathbf{x}$, and the oldest sample in $\mathbf{x}$ is discarded. The inner product
in (2.2) is then recomputed, and a new output sample is available. This structure,

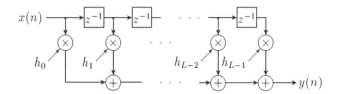

Figure 2.1. FIR filtering via direct form, or transversal structure.

also referred to as the transversal filter, is depicted in Fig. 2.1, where the operators z^{-1} are one-sample delays.

Either in hardware or software, the transversal filter is straightforward to implement, and it is such a basic processing unit that many digital signal processing integrated circuits have their hardware optimized to perform direct form FIR filtering at the fastest possible speed (generally one tap per instruction cycle) [2]. The disadvantage of the direct form is that, in general, L multiplications and $L - 1$ additions must be performed for each output sample. In applications such as acoustic echo cancelation, where filter lengths in excess of one thousand are employed, the computational effort of the direct form may be prohibitive.

The basic idea behind faster FIR filtering is to use the circular convolution property of the discrete Fourier transform (DFT) [1]. Given two vectors $\mathbf{x}$ and $\mathbf{h}$ of length M, with DFTs $\mathbf{X}$ and $\mathbf{H}$, respectively, we compute a vector $\mathbf{U}$ by

$$\mathbf{U} \equiv \mathbf{X} \times \mathbf{H} \tag{2.3}$$

where $\times$ denotes the Hadamard (element-by-element) product. Then, the inverse DFT of $\mathbf{U}$ is related to $\mathbf{x}$ and $\mathbf{h}$ by

$$\mathbf{u} = \mathbf{x} \circledast \mathbf{h}$$

where the operator $\circledast$ denotes circular convolution, which can also be written as

$$u(n) = \sum_{m=0}^{M-1} h(m)x((n - m) \bmod M) \tag{2.4}$$

where $i \bmod M$ denotes the remainder of the division of i by M. Note that the name circular convolution is related to the circular shift property of the mod operator; for example, $-1 \bmod M = M - 1$.

The circular convolution in (2.4) can be used to compute the linear convolution in (2.1). One way to accomplish that is to use the overlap-add procedure [1, 3], in which we overlap the results of circular convolutions to obtain the linear convolution. Another approach is the overlap-save, which is quite similar to the overlap-add. More details on the overlap-save procedure can be found in [1].

In order to see how the overlap-add procedure works, let us define two sequences of length M by overlap-add

$$h_r(n) = \begin{cases} h(n), & 0 \leq n \leq L - 1 \\ 0, & L \leq n \leq M - 1 \end{cases} \tag{2.5}$$

and

$$x_r(n) = \begin{cases} x(rL + n), & 0 \leq n \leq L - 1 \\ 0, & L \leq n \leq M - 1 \end{cases} \tag{2.6}$$

In the above definitions, we have assumed that $M = 2L$ (we can always choose L and M in order to satisfy this relationship). Thus, $h_r(n)$ is a block containing the L samples of $h(n)$, whereas $x_r(n)$ is a block containing L samples from $x(n)$, starting at $x(rL + n)$. Both blocks are padded with L zeros.

Assuming that the DFT of $x_r(n)$ and $h_r(n)$ are multiplied according to (2.3), and calling $u_r(n)$ the resulting sequence, we have, from (2.4)

$$u_r(n) = \sum_{m=0}^{M-1} h_r(m)x_r((n - m) \bmod M) \tag{2.7}$$

From the above equation and the definitions in (2.5) and (2.6), we can write

$$u_r(n) = \sum_{m=0}^{n} h_r(m)x_r(n - m) \, , \quad 0 \leq n \leq L - 1$$

and

$$u_r(n) = \sum_{m=n-L+1}^{L-1} h_r(m)x_r(n - m) \, , \quad L \leq n \leq M - 2$$

with $u_r(M - 1) = 0$.

Let us now overlap and add two consecutive blocks of $u_r(n)$, in the form

$$y(n) \equiv u_r(n \bmod L) + u_{r-1}(n \bmod L + L) , \quad rL \leq n \leq rL + L - 1 \qquad (2.8)$$

Using (2.6), we see that

$$x_r(n \bmod L - m) = x(n - m) , \quad 0 \leq m \leq n \bmod L$$

and

$$x_{r-1}(n \bmod L + L - m) = x(n - m) , \quad n \bmod L - L + 1 \leq m \leq L - 1$$

It is important to stress that these equations are only valid because we have padded $h_r(n)$ and $x_r(n)$ with zeros and taken length-M DFTs. Using the two equations above and also

$$h_r(m) = h(m) , \quad 0 \leq m \leq L - 1$$

we can write (2.8) in the form

$$y(n) = \sum_{m=0}^{L-1} h(m)x(n - m)$$

which is precisely the linear convolution that we originally wanted to implement in (2.1).

A graphical representation of the overlap-add procedure is shown in Fig. 2.2, where the output sequence $y(n)$ is formed by overlapping and adding the $u_r(n)$ blocks. Assuming that L is a power of two (we can always pad the original $h(n)$ with zeros in order to meet this), M is also a power of two, and so the popular radix-two or mixed-radix discrete Fourier transforms can be computed by the popular radix-two or mixed-radix fast Fourier transform (FFT), which are discussed in more detail in Section 2.5. In fact, as long as M is highly composite, we can use a fast DFT algorithm [4]

By replacing the direct-form linear convolution in (2.1) by the overlap-add procedure we have to perform, for each block of L output samples, one direct DFT (the DFT of $h(n)$ can be precomputed and stored), M complex multiplications,

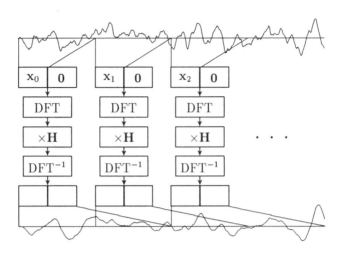

Figure 2.2. FIR filtering via overlap-add procedure. Other transforms can be used instead of the DFT.

one inverse DFT and $L - 1$ additions (because the last sample of $u_r(n)$ is always zero). In fact, due the complex conjugate symmetry of the DFT for real sequences [see (1.17)], only $L - 1$ complex multiplications and two real multiplications are needed to perform the product in (2.3). The computation of a length-M DFT of a real sequence by means of a FFT takes approximately $2M \log_2 M$ real operations when M is a power of two (the precise numbers are given in Section 2.5), and so the computational complexity of FIR filtering with the DFT will grow proportionally to the logarithm of the filter length. With direct-form filtering the computational effort grows proportionally to the filter length. As an example, if the filter length is $L = 256$, the direct form implementation requires a total of 511 arithmetic operations per output sample. With the overlap-add procedure the computational load is only 62 arithmetic operations per output sample.

From the example discussed above, it is clear that when we want to perform FIR filtering with a long filter, we should use the DFT-based overlap-add procedure, because the savings in computations can be of an order of magnitude or more. This argument alone would be enough to justify the search for efficient FFT algorithms, a topic to which we will return in Section 2.5.

The DFT is not the only transform that can be used for circular convolution (and hence for FIR filtering). The discrete Hartley transform (DHT) also has a convolution property [5, 6]. Calling $\mathbf{X}$, $\mathbf{H}$, and $\mathbf{U}$ the DHTs of the vectors $\mathbf{x}$, $\mathbf{h}$, and $\mathbf{u}$, respectively, the circular convolution of (2.4) leads to the following Hartley domain relationship

$$U(k) = X(k)H^e(k) + X(M - k)H^o(k) \qquad (2.9)$$

where $X(k)$ denotes the kth element of $\mathbf{X}$ and $H^e(k)$ and $H^o(k)$ are the even and odd parts of $\mathbf{H}$, that is

$$H^e(k) \equiv [H(k) + H(M - k)]/2$$

and

$$H^o(k) \equiv [H(k) - H(M - k)]/2$$

In many signal processing applications, it is desired have a linear phase filter, and so $h(n)$ will have even symmetry. In this case, $H^o(k) = 0$ and (2.9) will reduce to a simple element-by-element product, exactly like the DFT convolution property.

Once the DHT is used to perform cyclic convolution, it could replace the DFT in the overlap-add algorithm for FIR filtering. Compared to the DFT-based overlap-add procedure, the DHT requires a small increase in the number of additions [6] (less than 1% for $M = 512$). However, the DHT has the advantage of being its own inverse, so only one subroutine is necessary for transform computation. This is not true for the DFT with real input data, as we will see in Section 2.5, where a direct DFT subroutine cannot be simply modified to compute the inverse DFT.

The discrete cosine transform (DCT) can also be used to approximate cyclic convolutions [7], but an exact FIR filtering procedure of the overlap-add kind cannot be easily performed with DCTs. Nevertheless, in the context of signal restoration or coding, where the reconstructed signal will necessarily be subject to errors, filtering in the DCT domain as part of a signal coding or restoration process may be useful. In fact, DCT-domain filtering has been successfully applied to the problem of progressive image transmission with DCT domain filtering based on the human visual system response [8]. More details can be found in [9].

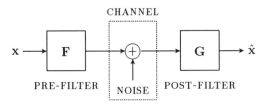

Figure 2.3. Block filtering for multichannel signal processing.

2.1.2   Multichannel Filtering

In biomedicine, data communications, geophysics, and other areas, it is common the presence of vector signals, that is, for a given time index n the signal of interest is composed of observations from many channels. Thus, a typical signal may be written as $\mathbf{x}(n)$, with $\mathbf{x}$ being an M-dimensional vector for a system with M channels.

A typical application in multichannel signal processing is the restoration of a transmitted signal that is corrupted by channel noise. A simple model for this problem is shown in Fig. 2.3, where the channel noise is modeled as additive [10]. The noise may be uncorrelated with the signal, as it is the case with thermal noise, or the two may be correlated, when we have crosstalk, for example [11].

The transmission (pre) and reception (post) filters in Fig. 2.3 are actually linear transformations associated with the matrices $\mathbf{F}$ and $\mathbf{G}$. Given the signal and noise autocorrelation matrices, it is relatively simple to derive a jointly-optimum pair of pre- and postfilters [10]. The optimum filters can be decomposed into diagonal factors and Karhunen-Loève transforms (KLTs). As we have already discussed in Chapter 1, the KLT can be approximated by the DCT without much loss of optimality, mainly for signals that can be represented by low-order AR models. In the context of block image processing, optimal pre- and postfilters based on the DCT and diagonal factors were used for image restoration, with good results [12].

2.1.3   Adaptive Filtering

In applications such as automatic equalization of unknown channels, system identification, and interference cancelling, we need to be able to adjust the filter coeffi-

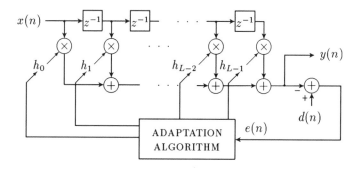

Figure 2.4. Adaptive FIR filter, transversal structure.

cients in Fig. 2.1 until the filter output matches a desired response. Assuming the filter is FIR, the transversal structure of Fig. 2.1 can be easily converted into an adaptive filter [13], as shown in Fig. 2.4.

The error signal $e(n)$, which is the difference between the desired signal $d(n)$ and the filter output $y(n)$, is the key information for the block "adaptation algorithm" in Fig. 2.4, which should try to minimize the error in some sense. When a least-mean-square criterion is used, a very simple adaptation rule can be derived [13]

$$\mathbf{h}_{n+1} = \mathbf{h}_n + 2\mu e(n)\,\mathbf{x}_n$$

where $\mathbf{h}$ is the L-dimensional vector containing the filter coefficients, and $\mathbf{x}_n$ is a running signal block, i.e.,

$$\mathbf{x}_n \equiv [x(n)\ x(n-1)\ \ldots\ x(n-L+1)]^T$$

The subscript n in $\mathbf{h}_n$ is important to denote the time variance of the filter coefficients. In fact, for every new incoming sample, a new set of filter coefficients is employed.

The parameter μ controls a fundamental trade-off between speed of convergence and stability. If μ is too small, it may take many thousands of iterations for $\mathbf{h}$ to converge, whereas if it is too large, the filter coefficients may diverge. The standard of comparison for the definition of what are small or large values of μ is the set

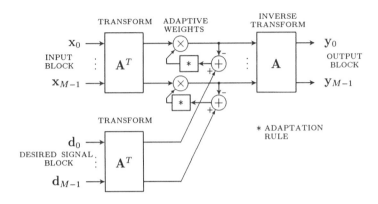

Figure 2.5. Frequency domain adaptive filter, structure FDAF-I.

of eigenvalues of the autocorrelation matrix of the input block $\mathbf{x}$. For stationary signals, the influence of μ on the learning characteristics of the adaptive filter can be found in [13]. For nonstationary signals, it is common to vary μ with time [14], because not only convergence is important, but also tracking, i.e., the ability of the adaptive filter to keep the lms error minimized under variations in the signal statistics.

We saw that in fixed FIR filtering a block-by-block processing approach with basis on a fast transform could significantly reduce the computational complexity. This is also true for adaptive filters, and in [15] it was suggested that the adaptive filter be implemented on a block-by-block basis, using the FFT. The structure of this frequency-domain adaptive filter (FDAF), which we will refer to as the FDAF-I, is shown in Fig. 2.5.

Besides coefficient adaptation, an important difference between the overlap-add procedure and the FDAF-I structure is that in the latter there is no zero padding of the input block, and there is no overlapping of the output blocks. Therefore, even if the complex weights are kept fixed, the structure of Fig. 2.5 does not correspond to a shift-invariant convolution. It is a periodically time-varying system, with a time-domain aliasing that leads to small discontinuities in $y(n)$ in the vicinity of the block boundaries (the blocking effects that we have discussed in Chapter 1). In applications were the filtered signal is already noisy, such as in data communications,

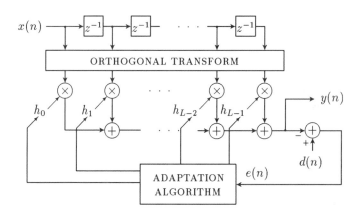

Figure 2.6. Frequency domain adaptive filter, structure FDAF-II.

this lack of shift invariance is of a minor concern. If necessary, the overlap-add or overlap-save approaches can easily be incorporated in the FDAF-I structure [16].

Although originally developed only with the aim of reducing the computational complexity, the FDAF has an important advantage over the transversal filter structure: it may have a much faster rate of convergence. In order to see this property, let us recall that the speed of convergence of a transversal LMS adaptive filter is controlled by the ratio $\lambda_{\max}/\lambda_{\min}$, where $\lambda_{\max}$ and $\lambda_{\min}$ are the largest and smallest eigenvalues of the covariance matrix of the input signal block [13]. The lower the input spectral flatness, the slower the convergence.

In [17] it was noted that the DFT coefficients are virtually uncorrelated among themselves and from block to block, since the DFT is an approximation to the KLT for large block sizes [18]. Therefore, each of the adaptive weights in Fig. 2.5 will converge much faster than the weights in the transversal filter structure. In [17] it was also proposed that the transform be computed on a sample-by-sample basis, as shown in Fig. 2.6.

With the structure in Fig. 2.6, which we will refer to as the FDAF-II, there is no inverse transform. In fact, the FDAF-II adaptive filter is a modification of the transversal filter in which the outputs of the tapped delay line (the cascade connection of delay units in Fig. 2.4) are first transformed before being combined by the adaptive weights. The purpose of the orthogonal transform in the FDAF-II structure of Fig. 2.6 is to act as a filter bank, so that its outputs are approximately uncorrelated. As demonstrated experimentally in [17], the convergence of the FDAF-II filter may be an order of magnitude faster than that of a transversal LMS filter. Although the Karhunen-Loève transform would be the optimal transform for the FDAF-II, because it would make the equivalent eigenvalue ratio of the transformed block $\lambda'_{\max}/\lambda'_{\min}$ equal to one, nearly optimal performance can be achieved with the DCT and other transforms [17].

In the FDAF-II structure, the LMS adaptation rule must be modified to consider the different variances of each transform coefficient. The transform-domain LMS update rule is [17]

$$[\mathbf{h}_{n+1}]_k = [\mathbf{h}_n]_k + 2\mu_k e_k(n) [\mathbf{x}_n]_k$$

Thus, there are M convergence control parameters, μ_0 to μ_{M-1}, one for each transform coefficient. Each μ_k is given by

$$\mu_k = \frac{\mu}{\sigma_k^2}$$

where σ_k^2 is the variance of the kth transform coefficient, which we saw how to compute in Section 1.2, and μ is an overall convergence control parameter.

One disadvantage of the FDAF-II structure is that the transform is computed on a sample-by-sample basis. Hence, the computational complexity of the FDAF-II structure is that of the transversal filter plus the complexity of the transform. If the whole block transform is computed for every new input sample, the complexity of the FDAF-II structure may be more than an order of magnitude higher that of the transversal filter, as noted in [19]. However, it is possible simply to update the transform for each new incoming sample, instead of recomputing all of it. With such a running DFT algorithm [20], the DFT can be updated with approximately $M/2$ complex multiplications and the same number of additions, and so the computational complexity of the FDAF-II is a little over three times that of the transversal structure. Therefore, the FDAF-II is better suited for applications requiring short filters [17]. If the DCT is used, running DCTs can also be easily implemented by making use of the shift properties of the DCT [21]. It is in fact relatively simple to derive efficient running algorithms for all transforms discussed in Chapter 1, except the KLT.

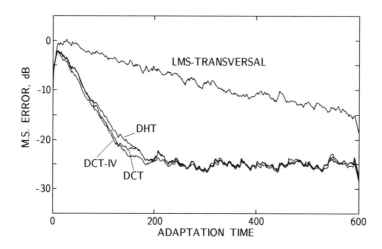

Figure 2.7. Error variance (in dB) versus adaptation time for the LMS-transversal and transform-domain LMS filters.

In order to have a better feeling for the performance of transform-domain adaptive filters, let us consider the following problem, which was simulated on a computer: a white Gaussian noise signal was transmitted through a linear channel with an all-pole transfer function

$$H(z) = \frac{1}{\left(1 - p_1 z^{-1}\right)\left(1 - p_2 z^{-1}\right)}$$

where the poles p_1, p_2 were located at $0.8e^{\pm j\pi/4}$. Besides this linear distortion, the channel added white Gaussian noise at a -30 dB level. The filter length was chosen to be $M = 8$, and we have assumed that the desired signal $d(n)$ was the original transmitted signal, i.e., an equalization problem. Note that the equalizer $H^{-1}(z)$ is a FIR filter of length 3, so we would expect the adaptive FIR filters to converge to $H^{-1}(z)$, with a steady-state error level close to the channel error level of -30 dB. The results of the computer simulation of this adaptive equalization problem are shown in Fig. 2.7, for the LMS-transversal filter and the FDAF-II based on the DHT, DCT, and DCT-IV. In each case, the largest possible value for the convergence control parameter μ within the stable range was used.

Each curve in Fig. 2.7 plots the estimated average error variance $E[e^2(n)]$ of 100 runs of the indicated adaptive filter. We see that all FDAF-II filters had similar performance, and they all converged to an error level of -25 dB in about 200 iterations, whereas the LMS transversal filter takes more than 1,000 iterations to converge. Note that the error was not expected to drop all the way to the channel noise level of -30 dB, due to the noise in the filter coefficients [13]. A slightly better steady-state error level could be achieved by using lower values of μ, but at the cost of longer convergence times.

In Chapter 7 we see that the performance of the FDAF filters can be improved by replacing the block transforms in Figs. 2.5 and 2.6 by lapped transforms. The main reason for that is the better decorrelation properties of LTs when compared to block transforms, as we will discuss in Chapters 4 and 5. Another important advantage of the use of LTs is that they allow the implementation of the FDAF-I structure without blocking effects.

2.2   Spectrum Estimation

In virtually any signal processing application, it is important to know the distribution of the energy of a signal in the frequency domain, i.e., its spectrum. We may be interested either in the frequency content of a single discrete-time sequence, or in the power density spectrum of a stochastic process, as we discussed briefly in Chapter 1. In the first case, we have a signal $x(n)$ and we want to estimate its Fourier transform

$$X(e^{j\omega}) = \sum_{n=-\infty}^{\infty} x(n)\, e^{-j\omega n}$$

and in the second case we want to estimate the Fourier transform of the autocorrelation sequence of the signal

$$S_{xx}(e^{j\omega}) = \sum_{n=-\infty}^{\infty} R_{xx}(n)\, e^{-j\omega n}$$

For real-world signals, we seldom know exactly $x(n)$ or $R_{xx}(n)$ for all n. Thus, exact computation of the spectrum is impossible, and so we must find an appropriate way of estimating it.

Unfortunately, there is no universal technique for spectrum estimation. A method that performs very well in providing a smooth spectrum estimate may be totally inapplicable for detecting sinusoids in noise, for example [22]. Therefore, we will only touch upon the subject here, with the purpose of highlighting how block transforms can be applied to spectrum estimation. There are many techniques that do not make use of transforms [22, 23], but we do not discuss them in this book.

One of the classical methods of spectrum estimation for random signals is the periodogram [22, 24], defined in (1.4) by

$$P_{xx}(e^{j\omega}) \equiv \frac{1}{M}|X(e^{j\omega})|^2$$

The above equation states that the power spectrum of a random signal can be estimated from the Fourier transform of one particular realization of the signal. As we have already discussed in Chapter 1, the expected value of the periodogram is the power spectrum $S_{xx}(e^{j\omega})$. However, the periodogram is generally noisy, so that the usual practice is to use the average of several periodograms over consecutive signal blocks. As long as the signal statistics do not change significantly during these blocks, the averaged periodogram may be a good enough spectrum estimate for most applications in which we are looking for a smooth estimate.

In order to compute the periodogram, though, we have to evaluate the Fourier transform of a signal block. However, $X(e^{j\omega})$ is a function of the continuous frequency variable ω, so we cannot compute $X(e^{j\omega})$ for every ω. Therefore, we generally use the DFT to obtain $X(e^{j\omega})$ on a grid of frequencies. The desired frequency resolution will determine the length M of the DFT, since the DFT frequencies are multiples of $2\pi/M$. Besides the discretization of the frequency scale, when we perform a DFT on a finite segment of data we are effectively windowing the signal by a rectangular window. Therefore, the DFT will not be a direct estimate of $X(e^{j\omega})$, but an estimate of $X(e^{j\omega})$ convolved with the window frequency response. To reduce the effects of the sidelobes of the window response there are many other windows, such as Hamming, Hann, and Kaiser, that are frequently employed instead of the rectangular window [22].

The major point that we wish to emphasize in this section is that the DFT is not the only transform that can be used to compute the periodogram. This is clear when we realize that the Fourier transform is actually a projection of the signal over a set of basis functions that are complex sinusoids. Therefore, any transform whose basis functions are sinusoids should serve for spectrum estimation. In fact,

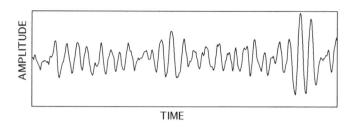

Figure 2.8. Random waveform used for the spectrum estimation test.

we recall from Chapter 1 that, whereas a length-M DFT has $M/2$ frequency bins in the interval $\omega \in [0, \pi]$, a length-M DCT has M frequency bins. Thus, the lack of phase information of the DCT is compensated by a better frequency resolution.

It is better to illustrate the ideas above by an example. Let us consider a random signal which is generated by passing white Gaussian noise through a second-order filter with transfer function

$$H(z) = \frac{1}{\left(1 - p_1 z^{-1}\right)\left(1 - p_2 z^{-1}\right)}$$

where the poles p_1, p_2 are located at $0.95e^{\pm j 7.5\pi/64}$. A sample waveform of the signal is shown in Fig. 2.8.

We have estimated the power spectrum of that signal by three similar implementations of the averaged periodogram: the first estimate was generated by averaging 16 periodograms computed with DFTs of length 128; the second was generated by averaging 32 periodograms computed with DCTs of length 64; and the third estimate was generated by averaging 32 periodograms computed with DCT-IVs of length 64. Note that a DHT-based periodogram is identical to a DFT-based periodogram [5]. In all cases, we have 64 frequency bins in the interval $\omega \in [0, \pi]$, and the total number of samples used was 2048.

In Fig. 2.9 we have the logarithmic power spectra, together with the exact power spectrum that was computed from (1.6). We see that all estimates are slightly noisy, but the DCT spectrum shows less noise at higher frequencies. The differences are small, though, so that the choice of the transform may be dictated by other factors. In transform coding, for example, where the DCT is generally employed,

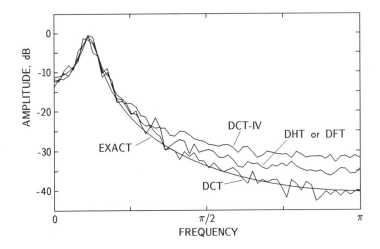

Figure 2.9. Spectrum estimates using the DFT, DCT, and DCT-IV.

blocks transformed with the DCT will have to be computed anyway, so if spectrum estimation is necessary, it may be done by DCT-based periodograms.

In Chapter 7 we discuss the use of lapped transforms in spectrum estimation, where we see that the intrinsic windowing associated with LTs helps to improve the quality of the estimates. An important observation will be that LT-based spectrum estimation has better frequency resolution than estimation based on block transforms for the same number of frequency bins.

2.3 Transform Coding

Digital signal coding is the process of representing a waveform as a sequence of bits, either for storage or transmission through a digital channel. One of the major goals of the designer of a digital coding system is to minimize the number of bits necessary to represent the signal, within a given fidelity criterion. This is because the cost of storage or transmission will inevitably increase with the number of bits. There are many techniques for digital signal coding, and virtually all of them are discussed in detail in [25].

The basic operation in signal coding is quantization, the process of converting analog (continuous) values into digital (discrete) values. The more discrete levels we assign to the output of a quantizer, the lower the error in approximating the original continuous variable, but also more bits will be required to encode the quantizer output. The simplest coding technique is pulse-code modulation (PCM), in which we periodically sample a continuous waveform, and quantize each sample with a prescribed number of bits. In log PCM, which is in widespread use in telephony, the discrete reconstruction levels at the quantizer output are nonuniformly distributed over the signal dynamic range. This allows for better reconstruction of nonstationary signals such as speech [25].

Whenever the incoming signal is strongly colored, i.e., when it has a low spectral flatness measure ($\gamma_x^2 \ll 1$), PCM is not an effective coding system. In fact, as we saw in Chapter 1, the rate distortion theory predicts that the minimum achievable distortion for a coder within a given rate is γ_x^2 times the PCM distortion for that same rate. Among the basic techniques that are able to perform closely to that bound are predictive coding, subband coding and transform coding. In this section we discuss the basics of transform coding, and in Chapter 3 we discuss subband coding and its close relationship to transform coding. Because predictive coding does not make use of transforms, it will not be discussed in this book (a detailed discussion of predictive coding can be found in [25]).

Transform coding was originally developed with the aim of extracting the redundancy that exists among the elements of a random vector. Such a correlation may exist either in a multichannel environment with cross-coupling among the channels [26], or in the case when the random vector is formed by a block of consecutive samples of a scalar signal [27]. The basic idea is to quantize the transformed vector instead of the original vector, as shown in Fig. 2.10. As we saw in Chapter 1, each transform coefficient has a different variance, even for a stationary input signal. Therefore, we could use fewer bits to quantize the coefficients with lower variances. This unequal bit distribution among the transform coefficient quantizers is the key point in transform coding, as we will see in the following.

Let us assume that the input vector $\mathbf{x}$ in Fig. 2.10 is formed by samples of a stationary signal $x(n)$ with autocorrelation function $R_{xx}(m)$. We recall from our discussion of the Karhunen-Loève transform (KLT) in Chapter 1 that the autocovariance matrix of $\mathbf{x}$ is given by

$$[\mathbf{R_{xx}}]_{nk} = R_{xx}(n-k) = R_{xx}(k-n)$$

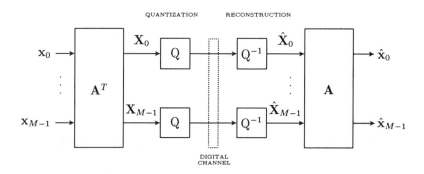

Figure 2.10. Block diagram of a transform coder.

and that the covariance of the transformed block $\mathbf{X}$ is

$$\mathbf{R_{XX}} = \mathbf{A}^T \, \mathbf{R_{xx}} \, \mathbf{A}$$

When we quantize each transform coefficient, i.e., each element of $\mathbf{X}$, we obtain an approximation of $\mathbf{X}$, which is referred to $\hat{\mathbf{X}}$ in Fig. 2.10. Assuming that ideal scalar quantizers are used, the mean-square error in the kth coefficient will be given, according to (1.10), by

$$E\{|[\hat{\mathbf{X}}]_k - [\mathbf{X}]_k|^2\} = 2^{-2B_k} \, E\{[\mathbf{X}]_k^2\} = 2^{-2B_k} \, [\mathbf{R_{XX}}]_{kk} = 2^{-2B_k} \, \sigma_k^2$$

where B_k is the number of bits used to quantize the kth coefficient, and σ_k^2 is the variance of that coefficient. Thus, the total mse ξ is given by

$$\xi = \sum_{k=0}^{M-1} 2^{-2B_k} \, \sigma_k^2 \tag{2.10}$$

where M is the block length.

At this point, we reach the classical bit allocation problem [25]. If we want to use an average rate of R bits per sample, we have a total of MR bits available for each block, that is

$$\sum_{k=0}^{M-1} B_k = MR \tag{2.11}$$

The problem then is how to assign the B_k bits such that ξ is minimized in (2.10). This is a simple constrained optimization problem, whose solution is the well-known log-variance rule [25–28]

$$B_k = \alpha + \frac{1}{2} \log_2 \frac{\sigma_k^2}{\sigma_{\text{GM}}^2} \qquad (2.12)$$

where σ_{GM}^2 is the geometric mean of the transform coefficient variances, i.e.,

$$\sigma_{\text{GM}}^2 \equiv \left(\prod_{k=0}^{M-1} \sigma_k^2 \right)^{1/M}$$

The parameter α in (2.12) is a Lagrange multiplier that should be chosen such that (2.11) is satisfied. Note that, if (2.12) predicts a negative number for a particular k, then the corresponding B_k should be set to zero. This clipping does not imply any loss in optimality [28]. In fact, for any resource allocation problem with a constraint on the total available amount of nonnegative resources and a concave cost function, the optimum solution is always obtaining by clipping, if necessary, the solution obtained without the nonnegativity constraint [29, Chapter 7].

When the average rate R is sufficiently large for all the B_k in (2.12) to be greater than zero, it is easy to show [25] that $\alpha = R$. In this case, the minimum mse is given by

$$\xi_{\min} = 2^{-2R} \sigma_{\text{GM}}^2$$

Coding with PCM at the same rate R leads to a mse

$$\xi_{\text{PCM}} = 2^{-2R} \sigma_x^2$$

where σ_x^2 is the variance of the input signal $x(n)$. We note that all diagonal elements of $\mathbf{R}_{xx}$ are equal to σ_x^2. Thus, because of the invariance of the trace under an orthogonal transformation [30], we have

$$\sigma_x^2 = \frac{1}{M} \sum_{k=0}^{M-1} \sigma_k^2$$

Therefore, the reduction in m.s.e. due to transform coding is given, for sufficiently high bit rates, by

$$G_{\mathrm{TC}} \equiv \frac{\xi_{\mathrm{PCM}}}{\xi_{\mathrm{min}}} = \frac{\sigma_x^2}{\sigma_{\mathrm{GM}}^2}$$

As we have seen in Chapter 1, the KLT will minimize the geometric mean of the transformed coefficient variances, and so it will lead to the maximum coding gain for any given block length M. However, the KLT has no fast algorithm and is signal dependent, so the question arises if wether a fast-computable transform can lead to a G_{TC} close to that of the KLT. The answer to that is fortunately positive. In fact, as we have note in Chapter 1, the DCT is asymptotically equivalent to the KLT for a first-order Markov signal with a high inter-sample correlation coefficient. Also, all sinusoidal orthogonal transforms are asymptotically equivalent to the KLT for any signal spectrum, when the block length M tends to infinity [18].

In Fig. 2.11 we have several plots of the transform coding gain G_{TC} for the KLT, DCT, DHT, DFT, and DCT-IV, for three input signal spectra. First, an AR(1) Markov process with an intersample correlation coefficient $\rho = 0.95$; second, an estimated spectrum from a voiced speech segment uttered by a male speaker; and third, an estimated spectrum from a voiced speech segment uttered by a female speaker. In all three cases and for all transforms, we have varied the block length M from 2 to 64 in powers of two[1].

We note that in virtually all cases the DCT performs as well as the KLT, with the DHT, DFT, and DCT-IV always having somewhat lower coding gains. For small block sizes, the DCT-IV is not useful, and for $M \geq 64$ all transforms have a G_{TC} within about 1 dB from the KLT for the simple first-order Markov model. With the more realistic spectra derived from speech signals, the DCT may have a coding gain more than 2 dB higher than the other transforms. Therefore, it is not without reason that the DCT is by far the most commonly used transform for signal coding [9, 25, 31].

The example discussed above raises an important issue in transform coding. The coder performance is certainly dependent on the input signal spectrum. Furthermore, changes in signal amplitude may further degrade the performance. This is because the reconstruction error in scalar quantization is highly sensitive to the

---

[1]In speech coding, block lengths larger than 64 are common.

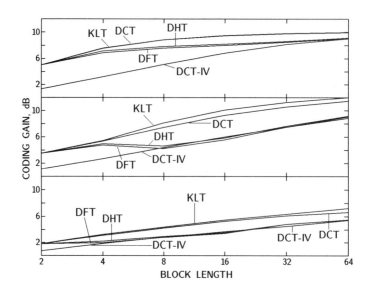

Figure 2.11. Coding gain (dB) versus block length M for several transforms. Top: for a first-order Markov process with $\rho = 0.95$; middle: for a male speech signal; bottom: for a female speech signal.

knowledge of the incoming signal variance. Any mismatch between the actual co-efficient variances and the variances that were used to design the quantizers will significantly increase the reconstruction error [25, Chapter 4]. Therefore, virtually any transform coder in practice must be adaptive.

Most adaptive transform coders are based on the block diagram of Fig. 2.12. The difference from Fig. 2.10 is that the quantizers are now adaptive, and are controlled by the block labeled "spectrum estimation." The purpose of that block is to estimate the variance distribution of the transform coefficients (which can always be viewed as a spectrum estimate, even if the transform is not the DFT, as we have discussed in the previous subsection). The design of this spectrum estimation block is certainly quite dependent on the input signal characteristics. It is also clear from Fig. 2.12 that the parameters that characterize the spectrum are side information that must also be sent to the receiver, so that the reconstruction levels of the quantizers are correctly determined. In practice, 20 to 40 percent of the available bit rate is assigned to this side information [9, 32].

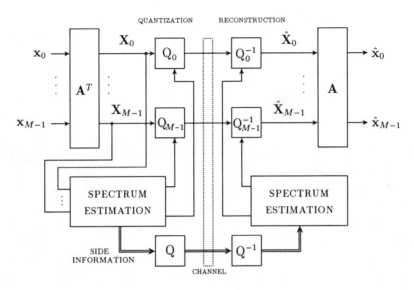

Figure 2.12. Block diagram of an adaptive transform coder.

In transform coding of images, one of the most popular implementations of the spectrum estimation block is the one in which all image blocks are assigned to one of several equally populated classes, according to some distinguishing feature, such as relative high-frequency content. Within each class, the variance of each transform coefficient is estimated by the average of the squares of that coefficient in all blocks of the class. This idea was originally suggested in [33] with four classes, and the results obtained were significantly better than with fixed quantization. In [34], this idea was generalized by modeling the input image as coming from a composite block source, and estimating the source parameters by a maximum likelihood (ML) procedure. With a large number of classes, the improvement in signal-to-noise ratio (SNR) over non-adaptive coding reported in [34] was in excess of 6 dB.

For transform coding of speech, it is hard to use the block classification strategy described above, because speech is not a finite length signal. Also, in order to avoid an excessive delay for real-time speech communication, we cannot wait for the arrival of a large number of blocks before we code the current block. Even if classes were built with basis on past blocks only, the number of classes would have to be too large, and only a moderate improvement in SNR could be expected [35].

Furthermore, speech signals cannot be considered stationary during intervals longer than 20 or 50 ms. Thus, statistics accumulated over a large number of samples are not helpful for adaptive speech coders.

After the issues raised above, we conclude that transform coefficient variance estimation in speech coding must be performed preferably with basis on the current block only. Three classical approaches toward this goal were suggested. In [35], the actual squared amplitudes of the transform coefficients are replaced by averages computed over a group of consecutive coefficients. The variance spectrum was recovered by linear interpolation. The side information is then composed by the averages of variances. A significant improvement in the quality of the spectrum estimation was obtained in [36], where an autoregressive model was used to represent a smoothed spectrum, and a pitch period estimator was included to capture the harmonic structure of voiced speech. Finally, in [37], a truncated cepstrum [1] was used to represent both the smooth formant structure and the pitch periodicity in the spectrum.

Further improvement on the performance of adaptive transform coders of images and speech has generally been obtained over the last decade by the use of vector quantization (VQ). The basic idea is to replace the M scalar quantizers in Figs. 2.10 and 2.12 by one or more vector quantizers. We believe that further discussion of VQ is beyond the scope of this book, so the interested reader is referred to [9, 25, 32].

Besides images and speech, other signals can also be successfully represented by transform coding. An example of seismic data compression at rates form 150 to 550 bits per second is reported in [38]. Whatever the signal, we can have a good idea of the usefulness of transform coding by computing the G_{TC} associated with the spectrum of that signal.

As we have already discussed in Section 1.3, the main motivation behind the development of lapped transforms was the removal of the major artifact generated by the transform coding technique: the blocking effects. These are discontinuities in the reconstructed signal in the vicinity of the block boundaries. As we see in Chapter 7, transform coding based on LTs is essentially free from blocking effects, with the added bonus of slightly higher coding gains.

2.4 Other Applications

There are other signal processing situations where orthogonal transforms can also be employed. One of them is in sampling rate conversion, either up-sampling (interpolation) or down-sampling (decimation). Traditionally, low-pass filters are used to avoid aliasing in decimation, and to remove the spectrum repetitions in interpolation. With the help of fast transforms, the computational complexity involved in sampling rate conversion can be significantly reduced. Not only the DFT [39], but also the DCT [9] and other sinusoidal transforms [40] can be used for efficient signal interpolation.

Speech scrambling is another area where transforms can be used. Imagine that in the adaptive transform coder of Fig. 2.10, the stream of bits corresponding to the quantized coefficients is pseudo-randomly permuted, with a different permutation being used for each block. An eavesdropper would have to find the correct permutation in a trial-and-error basis, and this would be almost impossible for large M. For analog channels, as is the case in standard telephony or radio communication, a simple scrambling technique is to resynthesize a block of speech from a pseudorandom permutation of its transform coefficients. This scrambled signal (unintelligible) would then be transmitted over the analog channel. At the receiver, the original signal would be recovered by block transformation, inverse permutation, and inverse transformation. When transform domain scrambling is combined with time domain block permutation, high levels of security can be achieved [41].

Other signal processing applications where transforms are useful include cepstrum analysis and simultaneous range-velocity determination in radar signal processing [42, Chapter 1]. In these areas, our feeling is that lapped transforms may not have significant advantages over block transforms.

2.5 Fast Algorithms

In this chapter we have seen that transforms have many applications in digital signal processing. In most cases, the advantages of using an orthogonal transform include the savings in computational effort due to fast algorithms. Since the development of the original FFT algorithm in [43], the field of fast algorithms for orthogonal transforms has experienced an enormous growth, and nowadays are highly optimized algorithms for all the commonly used orthogonal transforms. The purpose of this section is to present some of the most efficient algorithms for the transforms that we will use throughout this book, namely the DFT, DHT, DCT, and DCT-IV.

2.5.1   Discrete Fourier Transform

The computation of a length-M DFT of a signal block as a matrix multiplication according to (1.15) requires M^2 multiplications and approximately the same number of additions. In a classical paper by Cooley and Tukey [43], an FFT algorithm was presented for DFT computation with a computational effort proportional to $M \log_2 M$, for $M = 2^m$. For $M = 64$, for example, the computational effort is an order of magnitude lower than that of the direct form. It is not without reason, therefore, that the FFT represented a major contribution for spectrum estimation and signal filtering. It is interesting to note, however, that the same divide-and-conquer concept that is used in the FFT was employed in the computation of the DFT by Gauss in the early nineteenth century [44]. Since the original Cooley-Tukey algorithm, many FFT variations have been developed, including decimation in frequency or in time, decimations by factors of two, four, and eight. When the time or frequency decimation is two, for example, we have a radix-two algorithm. Fast algorithms for M not equal to a power of two have also been developed [4].

We could certainly spent a whole chapter (or more) if we went on describing all the various FFT algorithms. Instead, in most applications it is always possible to work with $M = 2^m$, and so algorithms based on radix-two or radix-four decompositions, or a combination of them, are quite useful. Therefore, we will restrict our discussion here to one of the best FFT algorithms for $M = 2^m$, the "split-radix" FFT (SRFFT) [6, 45, 46]. When we say best we mean that the SRFFT has the lowest number of multiplications and additions among all FFT algorithms known to date, and it is also easy to program in a computer. Furthermore, the number of multiplications required by the SRFFT is close to the theoretical minimum [47], and even when we take into account the additions, there is little room (if any) for further reductions in computational complexity beyond the SRFFT [47].

From the definition of the orthogonal DFT in (1.15), we can write a length-M DFT in the form

$$X_k = \sum_{n=0}^{M-1} x_n W_M^{kn} \qquad (2.13)$$

where X_k is the kth transform coefficient, $k = 0, 1, \ldots, M - 1$, and x_n the nth sample in the input block. We have changed notation from $x(n)$ and $X(k)$ to x_n and X_k, respectively, to simplify the forthcoming expressions. As usual, W_M is the Mth root of unity, that is,

$$W_M^r \equiv \exp\left(-j\frac{2\pi r}{M}\right)$$

We have dropped the $\sqrt{1/M}$ normalization factor because it will just determine the dynamic range of the DFT coefficients. If necessary, the DFT coefficients may be scaled after the FFT calculation.

The SRFFT is a mixed-radix approach. In its decimation-in-frequency (DIF) version, the frequency index k is split into three subsets, in the form [46]

$$X_{2k} = \sum_{n=0}^{M/2-1} (x_n + x_{n+M/2})W_{M/2}^{kn}$$

$$X_{4k+1} = \sum_{n=0}^{M/4-1} [(x_n - x_{n+M/2}) - j(x_{n+M/4} - x_{n+3M/4})]W_M^n W_{M/4}^{kn}$$

$$X_{4k+3} = \sum_{n=0}^{M/4-1} [(x_n - x_{n+M/2}) + j(x_{n+M/4} - x_{n+3M/4})]W_M^{3n} W_{M/4}^{kn} \qquad (2.14)$$

Thus, we see that the first sum in the above equation is a length-$M/2$ DFT, whereas the other two sums are length-$M/4$ DFTs. The input vectors for these shorter-length DFTs are easily computed from the original input sequence x_n. The above equations can be performed in place, and some of the multiplications factors are trivial. These simplifications are used in the SRFFT description below.

Split-Radix FFT Algorithm (SRFFT)

Step 1: Starting with an input vector composed of M complex numbers $[x_n]$, $n = 0, 1, \ldots, M - 1$, compute, for $n = 0, 1, \ldots, M/2 - 1$

$$x_n \;\leftarrow\; x_n + x_{n+M/2}$$
$$x_{n+M/2} \;\leftarrow\; x_n - x_{n+M/2}$$

Note that the two assignments above are the same step, and so in the second assignment the value of x_n before the first assignment should be used. In a computer program, this would require storing x_n in a temporary variable before performing the first assignment. This comment also applies to the following step.

Step 2: Compute, for $n = 0, 1, \ldots, M/4 - 1$

$$x_{n+M/2} \leftarrow x_{n+M/2} + j x_{n+3M/4}$$
$$x_{n+3M/4} \leftarrow x_{n+M/2} - j x_{n+3M/4}$$

Step 3: Compute, for $n = 1, 2, \ldots, M/4 - 1$,

$$x_{n+M/2} \leftarrow x_{n+M/2} W_M^n, \quad n \neq M/8$$
$$x_{5M/8} \leftarrow \sqrt{1/2}\,(1 - j)\,x_{5M/8}$$

Step 4: Compute, for $n = 1, 2, \ldots, M/4 - 1$,

$$x_{n+3M/4} \leftarrow x_{n+3M/4} W_M^{3n}, \quad n \neq M/8$$
$$x_{7M/8} \leftarrow -\sqrt{1/2}\,(1 + j)\,x_{7M/8}$$

Step 5: Compute

$$[x_0 \cdots x_{M/2-1}] \leftarrow \text{DFT}\{[x_0 \cdots x_{M/2-1}], M/2\}$$
$$[x_{M/2} \cdots x_{3M/4-1}] \leftarrow \text{DFT}\{[x_{M/2} \cdots x_{3M/4-1}], M/4\}$$
$$[x_{3M/4} \cdots x_{M-1}] \leftarrow \text{DFT}\{[x_{3M/4} \cdots x_{M-1}], M/4\}$$

where $\text{DFT}\{\cdot, L\}$ means a length-L DFT on the vector specified in the first argument.

Step 6: Data reordering

$$x_{2n} \leftarrow x_n, \quad n = 0, 1, \ldots, M/2 - 1$$
$$x_{4n+1} \leftarrow x_{n+M/2}, \quad n = 0, 1, \ldots, M/4 - 1$$
$$x_{4n+3} \leftarrow x_{n+3M/4}, \quad n = 0, 1, \ldots, M/4 - 1$$

An auxiliary vector of length M may be necessary to implement this step. The vector $[x_n]$ now contains, in order, the DFT coefficients.

A simplified block diagram of the above algorithm is shown in Fig. 2.13. We have not included all data paths in it for two reasons: first, because they can be inferred from Steps 1 to 6 above; second, because if we did the figure would be cluttered with crossing lines. Note that we have borrowed some concepts from signal flowgraphs:

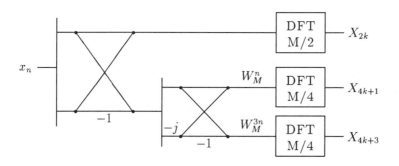

Figure 2.13. Simplified diagram of the split-radix FFT.

when a number is next to a path, it means that the signals on the path are multiplied by the number, and when two paths merge their signals are added. Note that we have not used arrows in the flowgraph because it is bidirectional. When data are entered from the left, a forward DFT is computed, and when the DFT coefficients are entered from the right, the inverse DFT is computed. Each line carries not a single coefficient, but a group of them. Finally, a vertical bar with one path in one side and two paths in the other means breaking a vector into two halves (or combining the two halves into a single vector, if processing goes from left to right).

Thus, Fig. 2.13 is not intended to be a rigorous flowgraph or block diagram of the SRFFT algorithm. The figure provides a pictorial representation of the steps described before. For example, Step 1 is a set of $M/2$ butterflies with ± 1 coefficients, and Step 2 is performed by $M/4$ ± 1 butterflies applied to the bottom half of the vector generated after Step 1.

The computational complexity of the SRFFT, measured as the total number of multiplications and additions, can be easily determined by counting the number of operations in Steps 1 to 6 above. Most of the operations are done in Steps 3 and 4, where complex multiplications are necessary. In most computers, a multiplication will take more time to execute than an addition, and then it is better to perform a complex multiplication as illustrated in Fig. 2.14, where three multiplications and three additions are used.

Counting the operations in Steps 1 to 6, we can write the complexity of a length-M SRFFT from of the complexities of SRFFTs of lengths $M/2$ and $M/4$.

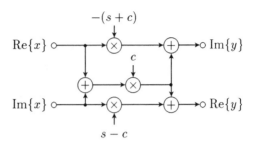

Figure 2.14. Complex multiplication $y = x\,(c - js)$, with c and s real, implemented with three multiplications and three additions.

Calling $\mu_C(M)$ and $\alpha_C(M)$ the number of real multiplications and real additions necessary to compute a length-M complex DFT via the SRFFT, and using the initial conditions $\mu_C(2) = \mu_C(4) = 0$, $\alpha_C(2) = 2$, and $\alpha_C(4) = 8$, we obtain the computational complexity of the SRFFT as [46]

$$\mu_C(M) = M(\log_2 M - 3) + 4$$
$$\alpha_C(M) = 3M(\log_2 M - 1) + 4$$

The SRFFT algorithm as described in Steps 1 to 6 above can be easily programmed in virtually any computer language. Furthermore, for those languages that allow recursion, the DFT computations in Step 5 can be performed by calling the SRFFT routine again. In order to stop the recursion, the SRFFT routine would use in-line code for the trivial cases $M = 2$ and $M = 4$ and would not call itself again. In such an implementation, the data reordering in Step 6 should be moved out of the recursive SRFFT module. All the data shuffling can be done in a single step after the recursive DFT computation. An example of a SRFFT program incorporating all of these ideas can be found in Appendix B.

As we have already noted above, the inverse DFT could be computed simply by running the flowgraph in Fig. 2.13 backwards. However, writing a computer program for this is not really necessary because (2.13) implies

$$jx_n^* = \frac{1}{M} \sum_{n=0}^{M-1} jX_k^* W_M^{kn}$$

which was first suggested in [48]. In most applications, the scale factor $1/M$ may not be necessary. Thus, an inverse DFT can be easily programmed simply by switching the real and imaginary parts of the input vector, computing a direct DFT, and switching the real and imaginary parts of the resulting vector, as shown in Appendix B.

When the input vector $[x_n]$ is real, the conjugate symmetry of the DFT coefficients, $X_{M-k} = X_k^*$, allows further simplification of the SRFFT. After Step 1 the vector $[x_n]$ is still real, and so the first DFT in Step 5 has real inputs. Furthermore, the X_{4k+3} coefficients can be obtained from the X_{4k+1} coefficients, using the conjugate symmetry. Thus, the third DFT in Step 5 need not computed, and the vector $[x_{3M/4} \cdots x_{M-1}]$ can be used to store the imaginary parts of the vector $[x_{M/2} \cdots x_{3M/4-1}]$. These considerations result in the following algorithm for real-input SRFFT, which we will refer to as the RSFFT.

Split-Radix FFT Algorithm for Real Inputs (RSFFT)

Step 1: Starting with an input vector of M real numbers, $[x_n]$, $n = 0, 1, \ldots, M-1$, compute, for $n = 0, 1, \ldots, M/2 - 1$

$$x_n \leftarrow x_n + x_{n+M/2}$$
$$x_{n+M/2} \leftarrow x_n - x_{n+M/2}$$

Step 2: For $n = 0, 1, \ldots, M/4 - 1$

$$x_{n+3M/4} \leftarrow -x_{n+3M/4}$$

After this step, we look at the lower half of $[x_n]$ as a $M/4$-element complex vector, which is split in two real vectors: the real part is in $[x_{M/2} \cdots x_{3M/4-1}]$, whereas the imaginary part is stored in $[x_{3M/4} \cdots x_{M-1}]$.

Step 3: Compute, for $n = 1, 2, \ldots, M/4 - 1$

$$x_{n+M/2} \leftarrow x_{n+M/2} W_M^n, \quad n \neq M/8$$
$$x_{5M/8} \leftarrow \sqrt{1/2}\,(1-j)\,x_{5M/8}$$

Note that for this step, we make use of the fact the $[x_{M/2} \cdots x_{3M/4-1}]$ is a complex vector, as noted in the previous step. The real part of the results should be stored in in $[x_{M/2} \cdots x_{3M/4-1}]$ and the imaginary part in $[x_{3M/4} \cdots x_{M-1}]$.

Step 4: Compute

$$[x_0 \cdots x_{M/2-1}] \;\leftarrow\; \mathrm{DFT}\{[x_0 \cdots x_{M/2-1}], M/2\}$$
$$[x_{M/2} \cdots x_{M-1}] \;\leftarrow\; \mathrm{DFT}\{[x_{M/2} \cdots x_{M-1}], M/4\}$$

Note that the first DFT above has a real argument, whereas the second is a complex DFT.

Step 5: Data reordering

$$x_{2n} \;\leftarrow\; x_n, \quad n = 0, 1, \ldots, M/4$$
$$x_{M/2+2n-1} \;\leftarrow\; x_{n+M/4}, \quad n = 1, 2, \ldots, M/4 - 1$$
$$x_{4n+1} \;\leftarrow\; x_{n+M/2}, \quad n = 0, 1, \ldots, M/8 - 1$$
$$x_{M/2+4n+1} \;\leftarrow\; x_{n+3M/4}, \quad n = 0, 1, \ldots, M/8 - 1$$
$$x_{4n+3} \;\leftarrow\; x_{3M/4-1-n}, \quad n = 0, 1, \ldots, M/8 - 1$$
$$x_{M/2+4n+3} \;\leftarrow\; -x_{M-1-n}, \quad n = 0, 1, \ldots, M/8 - 1$$

An auxiliary vector of length M may be necessary to implement this step. The vector $[x_n]$ now contains the DFT coefficients in the following order: the real parts of $[X_0 \cdots X_{M/2}]$ are stored in $[x_0 \cdots x_{M/2}]$, and the imaginary parts of $[X_1 \cdots X_{M/2-1}]$ are stored in $[x_{M/2+1} \cdots x_{M-1}]$. Note that the reordering assumed that a similar ordering was obtained form the DFT$\{\cdot, M/2\}$ in Step 4, and that the real and imaginary parts of the DFT$\{\cdot, M/4\}$ in Step 4 were stored in $[x_{M/2} \cdots x_{3M/4-1}]$ and $[x_{3M/4} \cdots x_{M-1}]$, respectively.

A block diagram of the above algorithm is shown in Fig. 2.15. As in Fig. 2.13, we have not used arrows in the flowgraph because it is bidirectional. When data is entered from the left, a forward DFT is computed, and when the DFT coefficients are entered from the right, the inverse DFT is computed.

A computer program implementing the RSFFT as described above is presented in Appendix B. The inverse RSFFT cannot be computed with a simple trick as in the case of the complex SRFFT. Thus, we have in Appendix B a program for the inverse RSFFT. The computational complexity of the real-valued length-M SRFFT, measured by the number of real multiplications $\mu_R(M)$ and real additions $\alpha_R(M)$, can be obtained from the algorithm described above as [46]

$$\mu_R(M) \;=\; \frac{M}{2}\left(\log_2 M - 3\right) + 2$$
$$\alpha_R(M) \;=\; \frac{M}{2}\left(3\log_2 M - 5\right) + 4$$

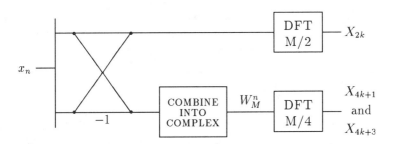

Figure 2.15. Simplified diagram of the split-radix FFT for real-valued inputs.

As a final note on the SRFFT computation, we see that W_M^r has the form $(c - js)$, where c and s are the cosine and sine of the angle $2\pi r/M$. Because the magnitudes of c and s are always less than one, the flowgraph in Fig. 2.14 has the desired property of not expanding much the dynamic range. In fact, if $\text{Re}\{x\}$ and $\text{Im}\{x\}$ have a maximum absolute value of Q, then $\text{Re}\{y\}$ and $\text{Im}\{y\}$ will have their maximum absolute values limited by $\sqrt{2}Q$. This leads to good numerical properties of the SRFFT algorithm; in a fixed-precision implementation (which is necessary in high-speed real-time applications) it is quite simple to avoid overflow within the SRFFT computations, and still keep the quantization noise within reasonable limits. We wanted to emphasize this point because not all fast transform algorithms have this desirable property.

2.5.2 Discrete Hartley Transform

The DHT is obtained by subtracting the imaginary part from the real part of the DFT sum in (2.13) (which corresponds to adding the real and imaginary parts of the basis functions, as we saw in Chapter 1). This leads to the following equation for computing the DHT

$$X_k = \sum_{n=0}^{M-1} x_n \operatorname{cas}\left(\frac{2\pi kn}{M}\right) \tag{2.15}$$

where $\mathrm{cas}(\theta) \equiv \cos(\theta) + \sin(\theta)$. As we have already seen in the previous section, the DHT can replace the DFT in any application where we deal with real signals. In fact, if we call F_k the kth DFT coefficient of $[x_n]$, then it is easy to see that

$$X_k = \mathrm{Re}\{F_k\} - \mathrm{Im}\{F_k\}$$

Therefore, a simple DHT algorithm would be to compute the RSFFT on the input vector and then use the equation above. In this way, the computational complexity of the DHT would be that of the RSFFT plus M additions. In fact, direct decompositions of the DHT such as the split-radix FHT (SRFHT) [49] cannot improve much on that; the SRFHT requires approximately $M/3$ extra additions when compared to the RSFFT. In [6] it was shown that by modifying slightly the twiddle factors (the W_M^n factors) in a decimation-in-time RSFFT, the RSFFT could be converted into an FHT routine with only two extra additions. However, it was noted in [6] that this gain in additions would not necessarily lead to a significant speed improvement over an SRFHT program. Thus, we will present here the SRFHT algorithm.

By breaking the frequency index k into three subsets, as it was done with the SRFFT, the DHT sum in (2.15) can be written in the form (after correcting a sign error from [49])

$$X_{2k} = \sum_{n=0}^{M/2-1} (x_n + x_{n+M/2}) \, \mathrm{cas}\left(\frac{2\pi kn}{M/2}\right)$$

$$X_{4k+1} = \sum_{n=0}^{M/4-1} \left[(x_n - x_{n+M/2} + x_{M/4-n} - x_{3M/4-n}) \cos\left(\frac{2\pi n}{M}\right) \right. $$
$$\left. + \; (x_{n+3M/4} - x_{n+M/4} + x_{M/2-n} - x_{M-n}) \sin\left(\frac{2\pi n}{M}\right) \right] \mathrm{cas}\left(\frac{2\pi kn}{M/4}\right)$$

$$X_{4k+3} = \sum_{n=0}^{M/4-1} \left[(x_n - x_{n+M/2} - x_{M/4-n} + x_{3M/4-n}) \cos\left(\frac{2\pi}{M} 3n\right) \right. $$
$$\left. - \; (x_{n+3M/4} - x_{n+M/4} - x_{M/2-n} + x_{M-n}) \sin\left(\frac{2\pi}{M} 3n\right) \right] \mathrm{cas}\left(\frac{2\pi kn}{M/4}\right)$$

Thus, a length-M DHT can be computed from one length-$M/2$ DHT, two length-$M/4$ DHTs and some extra operations. As it was the case with the SRFFT, the above equations can be performed in place, and some of the multiplications factors are trivial. These simplifications are used in the SRFHT description below.

Split-Radix Fast Hartley Transform Algorithm (SRFHT)

Step 1: Starting with an input vector of M real numbers, $[x_n]$, $n = 0, 1, \ldots, M-1$, compute, for $n = 0, 1, \ldots, M/2 - 1$

$$x_n \leftarrow x_n + x_{n+M/2}$$
$$x_{n+M/2} \leftarrow x_n - x_{n+M/2}$$

The comments made in the first step of the SRFFT algorithm also apply here.

Step 2: Compute, for $n = 0, 1, \ldots, M/8 - 1$

$$x_{n+M/2} \leftarrow x_{n+M/2} + x_{3M/4-n}$$
$$x_{3M/4-n} \leftarrow x_{n+M/2} - x_{3M/4-n}$$

Step 3: Compute, for $n = 1, 2, \ldots, M/8 - 1$

$$x_{n+3M/4} \leftarrow x_{M-n} + x_{n+3M/4}$$
$$x_{M-n} \leftarrow x_{M-n} - x_{n+3M/4}$$

Step 4: Set

$$x_{5M/8} \leftarrow \sqrt{2} x_{5M/8}$$
$$x_{7M/8} \leftarrow -\sqrt{2} x_{7M/8}$$

Step 5: Compute the butterflies, for $n = 1, 2, \ldots, M/8 - 1$

$$x_{n+M/2} \leftarrow \cos\left(\frac{2\pi n}{M}\right) x_{n+M/2} + \sin\left(\frac{2\pi n}{M}\right) x_{M-n}$$
$$x_{3M/4-n} \leftarrow \sin\left(\frac{2\pi n}{M}\right) x_{n+M/2} - \cos\left(\frac{2\pi n}{M}\right) x_{M-n}$$
$$x_{n+3M/4} \leftarrow \cos\left(\frac{6\pi n}{M}\right) x_{3M/4-n} - \sin\left(\frac{6\pi n}{M}\right) x_{n+3M/4}$$
$$x_{M-n} \leftarrow \sin\left(\frac{6\pi n}{M}\right) x_{3M/4-n} + \cos\left(\frac{6\pi n}{M}\right) x_{n+3M/4}$$

Note that this step actually computes two orthogonal butterflies at once, with output data shuffling. Thus, two temporary storage locations are needed. Furthermore, each orthogonal butterfly can be computed with three multiplications and three additions, in a structure similar to that of the complex multiplication of Fig. 2.14.

Step 6: Compute

$$
\begin{aligned}
[x_0 \cdots x_{M/2-1}] &\leftarrow \quad \text{DHT}\{[x_0 \cdots x_{M/2-1}], M/2\} \\
[x_{M/2} \cdots x_{3M/4-1}] &\leftarrow \quad \text{DHT}\{[x_{M/2} \cdots x_{3M/4-1}], M/4\} \\
[x_{3M/4} \cdots x_{M-1}] &\leftarrow \quad \text{DHT}\{[x_{3M/4} \cdots x_{M-1}], M/4\}
\end{aligned}
$$

where $\text{DHT}\{\cdot, L\}$ means a length-L DHT on the vector specified in the first argument.

Step 7: Data reordering

$$
\begin{aligned}
x_{2n} &\leftarrow \quad x_n, \quad n = 0, 1, \ldots, M/2 - 1 \\
x_{4n+1} &\leftarrow \quad x_{n+M/2}, \quad n = 0, 1, \ldots, M/4 - 1 \\
x_{4n+3} &\leftarrow \quad x_{n+3M/4}, \quad n = 0, 1, \ldots, M/4 - 1
\end{aligned}
$$

An auxiliary vector of length M may be necessary to implement this step. The vector $[x_n]$ now contains, in order, the DHT coefficients.

The simplified block diagram of the above algorithm is shown in Fig. 2.16. Although the flowgraph can be run from left to right to compute the inverse DHT, this is not necessary, since the DHT is its own inverse. A detailed view of each of the butterfly pairs that are needed in Step 5 is shown in Fig. 2.17. In the figure, we have used a standard flowgraph notation, with the multipliers c_i, s_i, c_{3i}, and s_{3i} being the cosines and sines defined in Step 5.

The computational complexity of the DHT can easily be evaluated. From the computational complexity algorithm described in Steps 1 to 7 above, we have

$$
\begin{aligned}
\mu(M) &= \mu(M/2) + 2\mu(M/4) + 6(M/8 - 1) + 2 \\
\alpha(M) &= \alpha(M/2) + 2\alpha(M/4) + M + 10(M/8 - 1) + 2
\end{aligned}
$$

With the initial conditions $\mu(2) = \mu(4) = 0$, $\alpha(2) = 2$, and $\alpha(4) = 8$, we obtain

$$
\begin{aligned}
\mu(M) &= \frac{M}{2}(\log_2 M - 3) + 2 \\
\alpha(M) &= \frac{3M}{2}\log_2 M - \frac{39M - 12(-1)^{\log_2 M}}{18} + 4
\end{aligned}
$$

A computer program implementing the SRFHT as described above is presented in Appendix B.

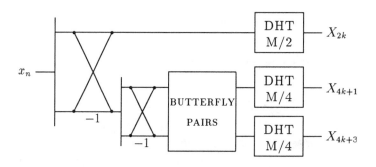

Figure 2.16. Simplified diagram of the split-radix FHT.

2.5.3 Discrete Cosine Transform

According to the definition of the DCT basis functions that we presented in Chapter 1, the DCT can be computed by the sum

$$X_k = c_k \sum_{n=0}^{M-1} x_n \cos\left[\left(n + \frac{1}{2}\right)\frac{k\pi}{M}\right] \tag{2.16}$$

where

$$c_k \equiv \begin{cases} 1/\sqrt{2} & \text{if } k = 0 \\ 1 & \text{otherwise} \end{cases} \tag{2.17}$$

Radix-two, radix-four, and split-radix algorithms for the DCT can also be developed. However, all such divide-and-conquer strategies, when used to obtain a length-M DCT from length-$M/2$ and length-$M/4$ DCTs, lead to additional operations that cannot be written as orthogonal butterflies [9, 50]. When $M = 128$, for example, there are four branch transmittances with factors greater than five (and one of them is approximately equal to 40), so it is quite difficult to achieve a good compromise between overflow and quantization noise. DCT algorithms with only orthogonal butterflies can be derived by direct factorization of the DCT matrix [51], but the resulting structures do not have a simple recursive form, so they are only useful for $M \leq 32$.

The most efficient algorithms for the DCT known to date were obtained by mapping the DCT into transforms of other types. Algorithms based on the DCT-IV [52],

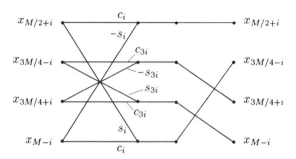

Figure 2.17. Structure of one of the butterfly pairs that are needed in the SRFHT. They can be implemented with six multiplications and six additions.

DFT (usually referred to as the FFCT [53]), and the DHT [54] have been developed, and they all have similar performances. In fact, the first two lead to exactly the same computational complexity, and are the best algorithms for the DCT known to date. Furthermore, it is virtually impossible to obtain any improvement on the number of multiplications and additions achieved by these two algorithms. The reasoning behind this argument is that if a better DCT could be found, it could be introduced in the FFCT algorithm to generate a new FFT algorithm that would be better than the SRFFT, which we know to be impossible, at least for $M < 128$ [47]. Thus, as in the case of the DFT and DHT, it is unlikely that DCT algorithms better than the FFCT and the one based on the DCT-IV could be developed.

The FFCT is based on taking the DFT of a rearranged input sequence. Given the input vector $[x_0 \cdots x_{M-1}]$, define a new vector $[y_n]$ by

$$
\begin{aligned}
y_n &= x_{2n} \\
y_{M-1-n} &= x_{2n+1}
\end{aligned}
$$

This mapping was first suggested in [55]. Now, let us call $[F_k]$ the DFT of $[x_n]$. Then, the DCT coefficients X_k can be obtained by [53]

$$
\begin{bmatrix} X_k \\ X_{M-k} \end{bmatrix} = \begin{bmatrix} \cos(\pi k/2M) & \sin(\pi k/2M) \\ \sin(\pi k/2M) & -\cos(\pi k/2M) \end{bmatrix} \begin{bmatrix} \mathrm{Re}\{F_k\} \\ \mathrm{Im}\{F_k\} \end{bmatrix}
$$

for $k = 1, 2, \ldots, M/2 - 1$. Furthermore, $X_0 = \sqrt{1/2}\, F_0$ and $X_{M/2} = \sqrt{1/2}\, F_{M/2}$.

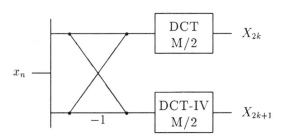

Figure 2.18. Simplified diagram of the fastest algorithm known to date for the computation of the DCT.

Computing each of the orthogonal butterflies above with three multiplications and three additions, and using the RSFFT for the computation of the DFT, we arrive at the following computational complexity for the FFCT

$$\mu(M) = \frac{M}{2} \log_2 M + 1$$

$$\alpha(M) = \frac{3M}{2} \log_2 M - M + 1$$

Note that in the value of $\mu(M)$ indicated above we have included the multiplication necessary to scale X_0 according to (2.17), which was not counted in [53].

A slightly simpler algorithm with exactly the same computational complexity of the FFCT can be obtained as suggested in [52]. Applying only one step of a decimation-in-frequency approach to the DCT in (2.16), we can write

$$X_{2k} = c_k \sum_{n=0}^{M/2-1} [x_n + x_{M-1-n}] \cos\left[\left(n + \frac{1}{2}\right) \frac{k\pi}{M/2}\right]$$

$$X_{2k+1} = \sum_{n=0}^{M/2-1} [x_n - x_{M-1-n}] \cos\left[\left(n + \frac{1}{2}\right)\left(k + \frac{1}{2}\right) \frac{\pi}{M/2}\right]$$

for $k = 0, 1, \ldots, M - 1$. Thus, the even-indexed DCT coefficients can be obtained from a length-$M/2$ DCT, whereas the sum that defines the odd-indexed coefficients corresponds to a length-$M/2$ DCT-IV. This is illustrated in Fig. 2.18.

Such a DCT decomposition into half-lengths DCT and DCT-IV is known for over a decade, since it was used in the direct DCT algorithms presented in [51, 56]. At

that time, however, the underlying DCT-IV was computed by a direct factorization
that was not very efficient. When we use the DCT-IV algorithm introduced in
the next subsection, though, the DCT algorithm of Fig. 2.18 has the following
computational complexity

$$\mu(M) = \mu(M/2) + \mu^{\text{DCT-IV}}(M/2) = \frac{M}{2} \log_2 M + 1$$
$$\alpha(M) = \alpha(M/2) + \alpha^{\text{DCT-IV}}(M/2) + M = \frac{3M}{2} \log_2 M - M + 1$$

Therefore, we arrive at precisely the same computational complexity of the FFCT.
However, the DCT-IV-based algorithm has the small advantage that only half of
the input vector must be rearranged (inside the DCT-IV algorithm, as we will see
in the next subsection), whereas with the FFCT the whole input vector must be
rearranged.

As a final comment on DCT computation, there are many applications where
including any scaling factor in *all* transform coefficients is irrelevant (e.g., in transform coding), as we have already discussed. We have emphasized the word "all"
because a different scaling factor for each transform coefficient could not be used
in applications such as transform coding, since the lack of a common normalization
factor would prevent us from using the log-variance bit allocation rule in (2.12).
If we multiply all coefficients by $\sqrt{2}$, for example, as suggested in [57], we could
save the two multiplications by $\sqrt{1/2}$ in the FFCT (the SRFFT complexity would
not be changed, since the $\sqrt{2}$ scaling could be incorporated in the twiddle factors
at no computational cost). Also, in the DCT-IV-based algorithm, we would have
as an initial condition $\mu(2) = 0$, and this would also save precisely two multiplications (the DCT-IV algorithm can also be slightly modified to accommodate the $\sqrt{2}$
scaling at no cost).

We conclude this subsection by noting that the DCT-IV based algorithm for the
computation of the DCT, with the $\sqrt{2}$ scaling of all transform coefficients, would
have exactly the same computational complexity as the optimal DCT algorithms
presented in [57]. However, structures only for $M = 8$ and $M = 16$ were presented
in [57], whereas the DCT-IV based algorithm can be easily implemented for all M.

2.5.4 Type-IV Discrete Cosine Transform

The DCT-IV is obtained, as we saw in Chapter 1, by shifting the frequencies of the DCT basis functions by $\pi/2M$. Thus, the DCT-IV of an input vector can be written as the sum

$$X_k = \sum_{n=0}^{M-1} x_n \cos\left[\left(n+\frac{1}{2}\right)\left(k+\frac{1}{2}\right)\frac{\pi}{M}\right] \tag{2.18}$$

where we have omitted the $\sqrt{2/M}$ normalization factor for simplicity. As we have seen earlier in this chapter, the DCT-IV can be used in spectrum estimation and adaptive filtering with good results. It is not appropriate for transform coding, though, due to the lack of direct dc information (note that none of the DCT-IV basis functions is a constant, so the signal dc component is spread among the basis functions).

Throughout this book, we will use the DCT-IV mainly as a building block for fast algorithms for the DCT and lapped transforms because the DCT-IV can be very efficiently computed, as we will see in the following. The first attempts to derive a fast algorithm for the DCT-IV were presented in [52, 56] by a direct factorization of the DCT-IV matrix, but with no recursive structure. In [58] a recursive DCT-IV algorithm was developed, based on the computation of a DCT-IV from two DCTs of half length. However, two butterfly stages were necessary to obtain such a decomposition, along with data reordering.

A simple but effective DCT-IV algorithm was unknowingly derived in [59, 60] as part of fast algorithms for a class of lapped transforms. The key idea is the same data shuffling that is the basis of the FFCT. Given the input vector $[x_0 \cdots x_{M-1}]$, define two new vectors $[u_n]$ and $[v_n]$ by

$$u_n = x_{2n}$$
$$v_n = x_{M-1-2n}$$

Then, (2.18) can be written as

$$X_k = \sum_{n=0}^{M/2-1} u_n \cos\left[\left(2n+\frac{1}{2}\right)\left(k+\frac{1}{2}\right)\frac{\pi}{M}\right]$$
$$+ \sum_{n=0}^{M/2-1} v_n \cos\left[\left(M-1-2n+\frac{1}{2}\right)\left(k+\frac{1}{2}\right)\frac{\pi}{M}\right]$$

which is equivalent to

$$X_{2k} = \text{Re}\left\{W_{2M}^k \sum_{n=0}^{M/2-1}(u_n + jv_n)W_{8M}^{4n+1}W_{M/2}^{kn}\right\}$$

$$X_{M-1-2k} = -\text{Im}\left\{W_{2M}^k \sum_{n=0}^{M/2-1}(u_n + jv_n)W_{8M}^{4n+1}W_{M/2}^{kn}\right\}$$

In the above equation, we see that $[X_k]$ can be obtained form a length-$M/2$ DFT of $(u_n + jv_n)W_{8M}^{4n+1}$. The corresponding in-place algorithm is presented below.

Fast DCT-IV Algorithm using the FFT

Step 1: Starting with an input vector of M real numbers, $[x_n]$, $n = 0, 1, \ldots, M-1$, rearrange the input sequence according to

$$x_n \leftarrow x_{2n}$$
$$x_{n+M/2} \leftarrow x_{M-1-2n}$$

for $n = 0, 1, \ldots, M/2 - 1$.

Step 2: Compute, for $n = 0, 1, \ldots, M/2 - 1$

$$x_n \leftarrow \text{Re}\left\{[x_n + jx_{n+M/2}]\exp\left[-j\left(n + \frac{1}{4}\right)\frac{\pi}{M}\right]\right\}$$

$$x_{n+M/2} \leftarrow \text{Im}\left\{[x_n + jx_{n+M/2}]\exp\left[-j\left(n + \frac{1}{4}\right)\frac{\pi}{M}\right]\right\}$$

After this step, we have a M-element complex vector, which is split in two real vectors: the real part is in $[x_0 \cdots x_{M/2-1}]$, and the imaginary part in stored in $[x_{M/2} \cdots x_{M-1}]$.

Step 3: Compute

$$[x_0 \cdots x_{M-1}] \leftarrow \text{DFT}\{[x_0 \cdots x_{M-1}], M/2\}$$

In this step, we use a length-$M/2$ to compute the DFT of $[x_n]$, still assuming that the first half of the vector contains the real part, and the second half contains the imaginary part, on both input and output.

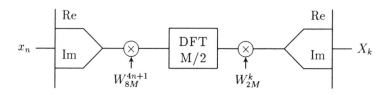

Figure 2.19. Simplified diagram of the fast DCT-IV algorithm.

Step 4: Compute

$$x_n \;\leftarrow\; x_n \exp\left(-j\frac{n\pi}{M}\right), \quad n = 1, 2, \ldots, M/4 - 1, M/4 + 1, \ldots, M/2 - 1$$

$$x_{M/4} \;=\; \sqrt{2}\,(1 - j)\, x_{M/4}$$

In this step, $[x_n]$ is still a complex vector, as noted above.

Step 5: Data reordering

$$x_{2n} \;\leftarrow\; x_n, \quad n = 0, 1, \ldots, M/2 - 1$$

$$x_{M-1-2n} \;\leftarrow\; -x_{n+M/2}, \quad n = 0, 1, \ldots, M/2 - 1$$

After this step, the output vector is again interpreted as a real vector, and contains, in order, the DCT-IV coefficients.

A block diagram of the above algorithm is shown in Fig. 2.19. The computational effort is that of one length-$M/2$ complex DFT plus $M/2$ complex multiplications in Step 2, plus $(M/2 - 2)$ complex multiplications and one multiplication by $\sqrt{2}(1 - j)$ in Step 4. Assuming that the DFT in Step 3 is computed by means of a complex SRFFT, and that the complex multiplications are performed according to Fig. 2.14, the computational complexity of the DCT-IV algorithm presented above is

$$\mu(M) \;=\; \frac{M}{2}\log_2 M + M$$

$$\alpha(M) \;=\; \frac{3M}{2}\log_2 M$$

Note that this is the lowest achievable complexity for the DCT-IV. If any better DCT-IV could be developed, it could be used in Fig. 2.18 to improve on the best known DCT algorithm, which in turn could be used in the FFCT to improve on the RSFFT, which is not possible. A computer program implementing the fast DCT-IV algorithm of Fig. 2.19 is presented in Appendix B.

LENGTH	DFT COMPLEX	DFT REAL	DHT	DCT	DCT-IV
2	0	0	0	2	3
4	0	0	0	5	8
8	4	2	2	13	20
16	20	10	10	33	48
32	68	34	34	81	112
64	196	98	98	193	256
128	516	258	258	449	576
256	1,284	642	642	1,025	1,280
512	3,076	1,538	1,538	2,305	2,816
1,024	7,172	3,586	3,586	5,121	6,144
2,048	16,388	8,194	8,194	11,265	13,312
4,096	36,868	18,434	18,434	24,577	28,672

Table 2.1. Number of multiplications required by the fast algorithms.

2.5.5　Computational Complexities

This has been a somewhat long section because of our goal of presenting in detail the best known algorithms for the DFT, DHT, DCT, and DCT-IV. We believe that this has been worthwhile because we need those fast algorithms in order to implement the lapped transforms efficiently. Furthermore, since it is now very unlikely that better algorithms for these transforms could be developed (unless M is not a power of two), we believe that the algorithms presented above, together with the computer programs of Appendix B, could be useful to the reader for some time.

We finish our discussion of fast algorithms by summarizing the computational complexity of all the algorithms presented above in Tables 2.1 and 2.2. Note that in the number of multiplications for the DCT we have considered the algorithm with the scaling factor as defined in (2.17). By using the scaling factor $\sqrt{2}$ in all coefficients, as commented when we presented the DCT algorithm, we could save two multiplications.

As a rule of thumb, we can say that the DFT for real signals and the DHT have roughly the same computational complexity. The DCT consumes approximately 80 percent more operations, and the DCT-IV is about 15 percent more expensive

LENGTH	DFT COMPLEX	DFT REAL	DHT	DCT	DCT-IV
2	4	2	2	2	3
4	16	6	8	9	12
8	52	20	22	29	36
16	148	60	66	81	96
32	388	164	174	209	240
64	964	420	442	513	576
128	2,308	1,028	1,070	1,217	1,344
256	5,380	2,436	2,522	2,817	3,072
512	12,292	5,636	5,806	6,401	6,912
1,024	27,652	12,804	13,146	14,337	15,360
2,048	61,444	28,676	29,358	31,745	33,792
4,096	135,172	63,492	64,858	69,633	73,728

Table 2.2. Number of additions required by the fast algorithms.

to compute than the DCT. With these percentages in mind, we can quickly have an idea of the performance versus complexity trade-off for the various transforms in a given application.

2.6 Summary

We have discussed in this chapter the applications of block transforms and fast algorithms for their fast computation. In Section 2.1 we saw that the computational complexity of FIR filtering can be reduced by using a fast transform. In adaptive filtering, the transform domain structures lead to faster convergence, as we verified in an example of the FDAF-II performance. Spectrum estimation is another classic area where transforms can be employed. We saw in Section 2.2 that the not only the DFT, but also the other sinusoidal transforms can be used in spectrum estimation.

In signal coding, we discussed the advantages of transform coding over PCM (straight quantization) in Section 2.3. Other applications, such as speech scrambling, were briefly discussed in Section 2.4. In Section 2.5 we studied in detail the best known algorithms for the DFT, DHT, DCT, and DCT-IV. Each algorithm was described completely in a step-by-step approach, which can help the reader implement it in a particular machine. For general-purpose computers, we have included programs in Appendix B for all of the discussed algorithms. An important feature of the algorithms that we discussed is that their performance, in terms of number of arithmetic operations, is already very close to the best that can be theoretically achieved (except for very long transforms, e.g., $M > 128$). Thus, they will be useful for quite some time.

In all of the applications that use a block transform, a lapped transform could also be employed. The advantages of lapped transforms over traditional block transforms depend on the application. In signal coding, for example, LTs have higher coding gains and no blocking effects. In spectrum estimation, LTs may have a better frequency resolution. In adaptive filtering with the FDAF-I structure, LTs will also attenuate the blocking effects and lead to slightly faster convergence than block transforms. These points are made clearer in Chapter 7, where we discuss in detail the applications of lapped transforms.

References

[1] Oppenheim, A. V., and R. Schafer, *Discrete-Time Signal Processing*, Englewood Cliffs, NJ: Prentice-Hall, 1989.

[2] Quarmby, D., ed., *Signal Processor Chips*, Englewood Cliffs, NJ: Prentice-Hall, 1985.

[3] Ludeman, L. C., *Fundamentals of Digital Signal Processing*, New York: Harper & Row, 1986.

[4] Blahut, R. E., ed., *Fast Algorithms for Digital Signal Processing*, Reading, MA: Addison-Wesley, 1985.

[5] Bracewell, R. N., "Discrete Hartley transform," *J. Opt. Soc. Am.*, vol. 73, Dec. 1983, pp. 1832–1835.

[6] Duhamel, P., and M. Vetterli, "Improved Fourier and Hartley transform algorithms: application to cyclic convolution of real data," *IEEE Trans. Acoust., Speech, Signal Processing*, vol. ASSP-35, June 1987, pp. 818–824.

[7] Chitpraset, B., and K. R. Rao, "Discrete cosine transform filtering," *IEEE Intl. Conf. Acoust., Speech, Signal Processing*, Albuquerque, NM, pp. 1281–1284, Apr. 1990.

[8] Chitpraset, B., and K. R. Rao, "Human visual weighted progressive image transmission," *IEEE Trans. Commun.*, vol. 38, July 1990, pp. 1040–1044.

[9] Rao, K. R., and P. Yip, *Discrete Cosine Transform: Algorithms, Advantages, Applications*, New York: Academic Press, 1990.

[10] Malvar, H. S., and D. H. Staelin, "Optimal pre- and post-filters for multichannel signal processing," *IEEE Trans. Acoust., Speech, Signal Processing*, vol. 36, Feb. 1988, pp. 287–289.

[11] Honig, M. L., K. Steiglitz, and B. Gopinath, "Multichannel signal processing for data communications in the presence of crosstalk," *IEEE Trans. Commun.*, vol. 38, Apr. 1990, pp. 551–558.

[12] Malvar, H. S., "Optimal pre- and post-filters in noisy sampled-data systems," Tech. Rep. 519, Research Lab. Electronics, Mass. Inst. Tech., Cambridge, MA, Sep. 1986.

[13] Widrow, B., and S. D. Stearns, *Adaptive Signal Processing*, Englewood Cliffs, NJ: Prentice-Hall, 1985.

[14] Karni, S., and G. Zeng, "A new convergence factor for adaptive filters," *IEEE Trans. Circuits Syst.*, vol. 36, July 1989, pp. 1011–1012.

[15] Dentino, M., J. McCool, and B. Widrow, "Adaptive filtering in the frequency domain," *Proc. IEEE*, vol. 66, Dec. 1978, pp. 1658–1659.

[16] Lee, J. C., and C. K. Un, "Performance analysis of frequency-domain block LMS adaptive filters," *IEEE Trans. Circuits Syst.*, vol. 36, Feb. 1989, pp. 173–189.

[17] Narayan, S., A. M. Peterson, and M. J. Narasimha, "Transform domain LMS algorithm," *IEEE Trans. Acoust., Speech, Signal Processing*, vol. ASSP-31, June 1983, pp. 609–615.

[18] Pearl, J., "On coding and filtering stationary signals by discrete Fourier transforms," *IRE Trans. Inform. Theory*, vol. IT-19, March 1973, pp. 229–232.

[19] Lee, J. C., and C. K. Un, "Performance of transform-domain lms adaptive digital filters," *IEEE Trans. Acoust., Speech, Signal Processing*, vol. ASSP-34, June 1986, pp. 499–510.

[20] Hartley, R., and K. Welles II, "Recursive computation of the Fourier transform," *IEEE Intl. Symp. Circuits Syst.*, New Orleans, LA, pp. 1792–1795, May 1990.

[21] Yip, P., and K. R. Rao, "On the shift property of DCT's and DST's," *IEEE Trans. Acoust., Speech, Signal Processing*, vol. ASSP-35, March 1987, pp. 404–406.

[22] Jr., S. L. M., ed., *Digital Spectral Analysis*, Englewood Cliffs, NJ: Prentice-Hall, 1987.

[23] Gardner, W. A., *Statistical Spectral Analysis: A Nonprobabilistic Approach*, Englewood Cliffs, NJ: Prentice-Hall, 1988.

[24] Blackman, R. B., and J. W. Tukey, *The Measurement of Power Spectra*, New York: Dover, 1958.

[25] Jayant, N. S., and P. Noll, *Digital Coding of Waveforms*, Englewood Cliffs, NJ: Prentice-Hall, 1984.

[26] Kramer, H. P., and M. V. Mathews, "A linear coding for transmitting a set of correlated signals," *IRE Trans. Inform. Theory*, vol. IT-2, Sep. 1956, pp. 41–45.

[27] Huang, J. J. Y., and P. M. Schultheiss, "Block quantization of correlated gaussian random variables," *IEEE Trans. Commun. Syst.*, vol. CS-11, Sep. 1963, pp. 289–296.

[28] Segall, A., "Bit allocation and encoding for vector sources," *IRE Trans. Inform. Theory*, vol. IT-22, March 1976, pp. 162–169.

[29] Luenberger, D. G., *Optimization by Vector Space Methods*, New York: Wiley, 1969.

[30] Gantmacher, F. R., *The Theory of Matrices, Vol. I*, New York: Chelsea, 1977.

[31] Clarke, R. J., *Transform Coding of Images*, London: Academic Press, 1985.

[32] Netravali, A. N., and B. G. Haskell, *Digital Pictures: Representation and Compression*, New York: Plenum, 1989.

[33] Chen, W. H., and C. H. Smith, "Adaptive coding of monochrome and color images," *IEEE Trans. Commun.*, vol. COM-25, Nov. 1977, pp. 1285–1292.

[34] Wornell, G., and D. H. Staelin, "Transform image coding with a new family of models," *IEEE Intl. Conf. Acoust., Speech, Signal Processing*, New York, pp. 777–780, Apr. 1988.

[35] Zelinski, R., and P. Noll, "Adaptive transform coding of speech signals," *IEEE Trans. Acoust., Speech, Signal Processing*, vol. ASSP-25, Aug. 1977, pp. 299–309.

[36] Tribolet, J. M., and R. E. Crochiere, "Frequency domain coding of speech," *IEEE Trans. Acoust., Speech, Signal Processing*, vol. ASSP-27, Oct. 1979, pp. 512–530.

[37] Cox, R. V., and R. E. Crochiere, "Real-time simulation of adaptive transform coding," *IEEE Trans. Acoust., Speech, Signal Processing*, vol. ASSP-29, Apr. 1981, pp. 147–154.

[38] Spanias, A. S., S. B. Johnson, and S. D. Stearns, "Transform coding algorithms for seismic data compression," *IEEE Intl. Symp. Circuits Syst.*, New Orleans, LA, pp. 1573–1576, May 1990.

[39] Fraser, D., "Interpolation by the FFT revisited–an experimental investigation," *IEEE Trans. Acoust., Speech, Signal Processing*, vol. 37, May 1989, pp. 665–675.

[40] Wang, Z., "Interpolation using type I discrete cosine transform," *Electron. Lett.*, vol. 26, no. 15, July 1990, pp. 1170–1172.

[41] Del Re, E. R., R. Fantacci, G. Bresci, and D. Maffucci, "A new speech signal scrambling method for mobile radio applications," *Alta Frequenza*, vol. LVII, Feb.-Mar. 1988, pp. 133–138.

[42] Beauchamp, K. G., *Transforms for Engineers: A Guide to Signal Processing*, New York: Oxford University Press, 1987.

[43] Cooley, J. W., and J. W. Tukey, "An algorithm for the machine calculation of complex Fourier series," *Math. Computation*, vol. 19, Apr. 1965, pp. 297–301.

[44] Heideman, M., D. H. Johnson, and C. S. Burrus, "Gauss and the history of the fast Fourier transform," *IEEE ASSP Magazine*, vol. 1, Oct. 1984, pp. 14–21.

[45] Duhamel, P., and H. Hollman, "Split-radix FFT algorithm," *Electron. Lett.*, vol. 20, no. 1, Jan. 1984, pp. 14–16.

[46] Duhamel, P., "Implementation of 'split-radix' FFT algorithms for complex, real, and real-symmetric data," *IEEE Trans. Acoust., Speech, Signal Processing*, vol. ASSP-34, Apr. 1986, pp. 285–295.

[47] Duhamel, P., "Algorithms meeting the lower bounds on the multiplicative complexity of length-2^n DFT's and their connection with practical applications," *IEEE Trans. Acoust., Speech, Signal Processing*, vol. 38, Sep. 1990, pp. 1504–1511.

[48] Duhamel, P., B. Piron, and J. M. Etcheto, "On computing the inverse DFT," *IEEE Trans. Acoust., Speech, Signal Processing*, vol. 36, Feb. 1988, pp. 285–286.

[49] Sorensen, H. V., D. L. Jones, C. S. Burrus, and M. T. Heideman, "On computing the discrete Hartley transform," *IEEE Trans. Acoust., Speech, Signal Processing*, vol. ASSP-33, Oct. 1985, pp. 1231–1238.

[50] Lee, B. G., "A new algorithm to compute the discrete cosine transform," *IEEE Trans. Acoust., Speech, Signal Processing*, vol. ASSP-32, Dec. 1984, pp. 1243–1245.

[51] Wang, Z., "Reconsideration of fast computational algorithm for the discrete cosine transform," *IEEE Trans. Commun.*, vol. COM-31, Jan. 1983, pp. 121–123.

[52] Wang, Z., "Fast algorithms for the discrete W transform and for the discrete Fourier transform," *IEEE Trans. Acoust., Speech, Signal Processing*, vol. ASSP-32, Aug. 1984, pp. 803–816.

[53] Vetterli, M., and H. J. Nussbaumer, "Simple FFT and DCT algorithms with reduced number of operations," *Signal Processing*, vol. 6, July 1984, pp. 267–278.

[54] Malvar, H. S., "Fast computation of discrete cosine transform through fast Hartley transform," *Electron. Lett.*, vol. 22, no. 7, March 1986, pp. 352–353.

[55] Narasimha, M. J., and A. M. Peterson, "On the computation of the discrete cosine transform," *IEEE Trans. Commun.*, vol. COM-26, June 1978, pp. 934–936.

[56] Chen, W. H., C. H. Smith, and S. C. Fralick, "A fast computational algorithm for the discrete cosine transform," *IEEE Trans. Commun.*, vol. COM-25, Sep. 1977, pp. 1004–1009.

[57] Loeffler, C. A., A. Ligtenberg, and G. S. Moschytz, "Practical fast 1-D DCT algorithms with 11 multiplications," *IEEE Intl. Conf. Acoust., Speech, Signal Processing*, Glasgow, Scotland, pp. 988–991, May 1989.

[58] Wang, Z., "On computing the discrete Fourier and cosine transforms," *IEEE Trans. Acoust., Speech, Signal Processing*, vol. ASSP-33, Oct. 1985, pp. 1341–1344.

[59] Cramer, S., and R. Gluth, "Computationally efficient real-valued filter banks based on a modified O^2DFT," *Proc. European Signal Processing Conf.*, Barcelona, Spain, pp. 585–588, Sep. 1990.

[60] Duhamel, P., Y. Mahieux, and J. P. Petit, "A fast algorithm for the implementation of filter banks based on time domain aliasing cancelation," *IEEE Intl. Conf. Acoust., Speech, Signal Processing*, Toronto, Canada, pp. 2209–2212, May 1991.

Chapter 3

Signal Processing in Subbands

The purpose of this chapter is to cover the basic theory, design techniques, and applications of subband signal processing. The main idea is that in many applications it may be desirable to separate the incoming signal into several subband components, by means of band-pass filters, and then process each subband separately. This will lead to extra degrees of freedom, which can be used for better processing results in coding, for example. Such a subband decomposition must be done with care, however, because of two major issues: first, if we simply pass the incoming signal through a set of band-pass filters and use their outputs directly, the amount of information that we will have to process will grow proportionally to number of subbands; second, we must design the subband filters such that we will be able to recover the original signal from its subbands.

Solving the information growth problem is relatively simple because the output of each band-pass filter has a limited bandwidth. In an M-band decomposition with equal-bandwidth filters, for example, the subband signals would have a bandwidth approximately equal to π/M. Therefore, we could subsample each subband signal by a factor of M, that is, we could keep only one out of each M samples. In this way, the total effective sampling rate would be preserved, and there would be no information growth. However, this subsampling (also referred to as decimation, or down-sampling) would lead to aliasing [1]. Because of this aliasing, it would be more difficult to recover the original signal from its decimated subbands, the second major issue that we cited above.

Thus, we see that an efficient subband decomposition would be one in which the set of band-pass filters (the analysis filter bank) is such that their outputs can be decimated with a relatively small amount of aliasing. Furthermore, the effects of

---

the aliasing should be minimized by the set of band-pass filters that will be used to recover the original signal from its subbands (the synthesis filter bank). In fact, by proper design of the filters, it is possible to achieve not only total elimination of the aliasing effects, but also to obtain perfect reconstruction, that is, the recovered signal is an exact delayed replica of the original signal. Throughout this chapter, we will discuss such design techniques.

We will begin our discussion of subband signal processing with a review of the basic theory of multirate systems in Section 3.1. This is necessary because we want to decimate the subband signals. The use of these concepts in the analysis of filter banks will be discussed in Section 3.2. Next, we will address the problem of eliminating the aliasing components generated by the decimation of the subband signals. This goal can be attained with the so-called quadrature mirror filters, which we will discuss in detail in Section 3.3. Besides aliasing cancellation, it is possible to achieve perfect reconstruction of the signal from its subband components, as we will see in Section 3.4. We recall that block transforms were used in Chapter 2 for breaking a signal into its frequency components. Therefore, there must exist some strong connection between subband processing and block transforms. This connection will be the topic of Section 3.5, where the relationships with lapped transforms will also be discussed. Finally, in Section 3.6 we will review some of the applications of subband signal processing.

3.1   Multirate Signal Processing

We refer to a signal processing system as a multirate system when more than one sampling rate is used. This is the case in subband processing, because the subband signals are generally decimated. In virtually all cases, there is one fundamental sampling frequency, the one that was used to obtain the original incoming signal. Further sampling rate alterations are done in the digital domain, that is, by down- and up-sampling of discrete-time signals.

3.1.1   Decimation and Interpolation

One of the two basic processing elements for sampling rate alterations is the $M:1$ decimator with input $x(n)$ and output $y(m)$, which is defined by

$$y(m) = x(mM) \tag{3.1}$$

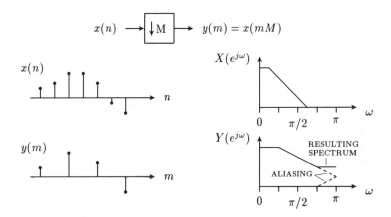

Figure 3.1. Decimation by a factor of M. Top: block diagram; bottom left: time domain operation, for $M = 2$; bottom right: equivalent spectral operation.

where we see that only one out of M input samples appears at the output. Note that we use the variable n to index the original sequence, and m to index the decimated sequence. This is to make clear that we have two clock counters: the clock for $y(m)$ runs at $1/M$th the rate of the clock for $x(n)$. In Fig. 3.1 we have the block diagram of the $M:1$ decimator, where we see that it generates a new sequence which contains every Mth sample of the original sequence.

The time domain operation in (3.1) corresponds to the following z-transform relationship [1, 2]

$$Y(z) = \frac{1}{M} \sum_{r=0}^{M-1} X(z^{1/M} W_M^r) \qquad (3.2)$$

where, as in Chapter 2, W_M is the Mth root of unity, $W_M \equiv e^{-j2\pi/M}$, with $j \equiv \sqrt{-1}$. In the Fourier domain, the above equation is equivalent to

$$Y(e^{j\omega}) = \frac{1}{M} \sum_{r=0}^{M-1} X(e^{j(\omega-2\pi r)/M})$$

We see that the spectrum of the output of the decimator if formed by the superposition of M replicas of the input spectrum expanded in frequency by a factor of M. If the input signal $x(n)$ has a bandwidth larger than π/M, there will be regions of overlapping of these spectral replicas. This phenomenon, illustrated in Fig. 3.1, is referred to as *aliasing* [1]. When aliasing occurs it is not possible to recover $X(e^{j\omega})$ from $Y(e^{j\omega})$, because the resulting spectrum in the aliasing region will not duplicate the original spectrum. Therefore, it is usual to precede a decimator by a low-pass or band-pass filter with a bandwidth π/M to minimize the effects of aliasing.

The other basic processing element in multirate signal processing is the $1:M$ interpolator with input $y(m)$ and output $x(n)$, which is defined by

$$x(n) = \begin{cases} y(n/M), & \text{if } n \bmod M = 0 \\ 0, & \text{otherwise} \end{cases} \tag{3.3}$$

Thus, the output sequence is formed by inserting $M - 1$ zero samples between each pair of input samples. As in the case of the decimator, we use different variables to index the sequences at the different sampling rates. Throughout this book, the interpolation factors will almost always be identical to the decimation factors. When this happens, we effectively work with only two sampling frequencies. Sequences at the same sampling rate as the input signal will be indexed by the variable n, and decimated sequences will be indexed by the variable m.

As is clear from (3.3), the interpolator generates a new sequence that is formed by the original samples of the input sequence padded by $M - 1$ zeros between each pair of original samples. The z-transforms of the input and output of the interpolator are related [1, 2] by

$$X(z) = Y(z^M) \tag{3.4}$$

or, in the Fourier domain,

$$X(e^{j\omega}) = Y(e^{j\omega M})$$

In Fig. 3.2 we have the block diagram of the $1:M$ interpolator. We see that the output spectrum is now formed by shrinking the original spectrum in frequency by a factor of M, and replicating this shrunk spectrum $M - 1$ times. Therefore, there is no aliasing, but the output spectrum $X(e^{j\omega})$ contains multiple replicas of $Y(e^{j\omega})$. This phenomenom, illustrated in Fig. 3.2, is usually referred to as *imaging*.

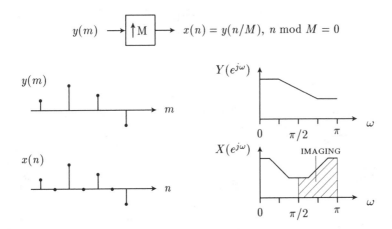

Figure 3.2. The $1:M$ interpolator. Top: block diagram; bottom left: time domain operation, for $M = 2$; bottom right: equivalent spectral operation.

In order to avoid imaging, it is usual to follow the interpolator with a low-pass or band-pass filter having bandwidth equal to π/M, so that only one of the spectral images will remain. When a low-pass filter is used, the equivalent time-domain operation is to replace the zero samples of $x(n)$ (which were inserted by the interpolator) by new samples values, in between their neighboring values. This is what we usually understand intuitively as interpolation. In terms of multirate signal processing, therefore, interpolation can be viewed as the cascade of an interpolator and a low-pass filter.

It is possible to alter the sampling rate by a noninteger factor simply by cascading a $1:L$ interpolator with a $M:1$ decimator [1]. The new sampling rate would then be equal to L/M times the original sampling rate. This idea is illustrated in Fig. 3.3, where the filter $H(z)$ should have a bandwidth of $\pi/\max\{M, L\}$. Note that $H(z)$ has a dual purpose: removing the imaging caused by the $1:L$ interpolator, and the aliasing caused by the $M:1$ decimator. In most cases $H(z)$ will be a low-pass filter, but a band-pass filter can also be used. If $L > M$, we have an overall sampling rate increase with the system in Fig. 3.3, and so no information will be lost. When $L < M$, some of the frequency components of $x(n)$ may be lost. Then, $y(m)$ will be a resampled and filtered version of $x(n)$.

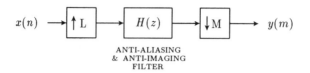

Figure 3.3. Sampling rate alteration by a rational factor L/M.

3.1.2   Cascade Connections

Another cascade connection of a decimator and an interpolator that will be useful in the following sections is the one depicted in Fig. 3.4. We first decimate the incoming signal by a factor of M, and then we bring it back to the original sampling rate with a $1:M$ interpolator. In this case, the input and output are at the same sampling rate, and this is why the independent variable is n for both. From the definitions in (3.1) and (3.3), we see that $y(n)$ and $x(n)$ in Fig. 3.4 are related by

$$y(n) = x(n)\, \delta_M(n) \;=\; \begin{cases} x(n), & \text{if } n \bmod M = 0 \\ 0, & \text{otherwise} \end{cases} \tag{3.5}$$

where the periodic sampling sequence $\delta_M(n)$ is defined by

$$\delta_M(n) = \sum_{r=-\infty}^{\infty} \delta(n + rM)$$

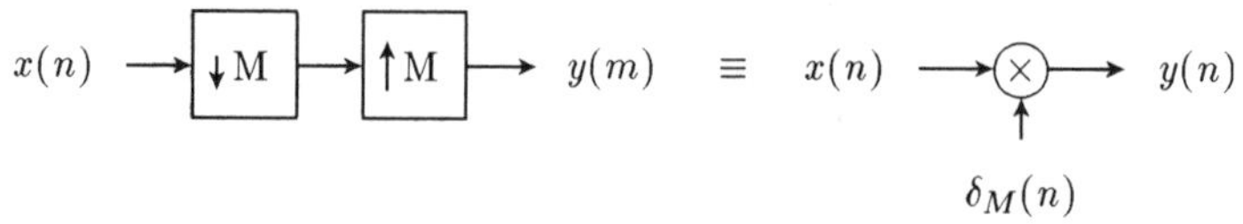

$$x(n) \longrightarrow \boxed{\downarrow M} \longrightarrow \boxed{\uparrow M} \longrightarrow y(m) \quad \equiv \quad x(n) \longrightarrow \otimes \longrightarrow y(n)$$
$$\delta_M(n)$$

Figure 3.4. Periodic sampling as a cascade of decimation and interpolation.

This representation of the cascade of a decimator and a interpolator as a periodic sampler has been used in the analysis of optimum pre- and postfilters [3, 4]. In terms of z-transforms, (3.5) can be written as

$$Y(z) = \frac{1}{M} \sum_{r=0}^{M-1} X(z W_M^r) \tag{3.6}$$

or, in the Fourier domain,

$$Y(e^{j\omega}) = \frac{1}{M} \sum_{r=0}^{M-1} X(e^{j(\omega - 2\pi r/M)})$$

Thus, $Y(e^{j\omega})$ contains aliased versions of the original spectrum $X(e^{j\omega})$, but at the same sampling frequency, and hence the same frequency scale.

Decimators and interpolators are linear but shift-variant operators. Therefore, in the cascade of a filter and a decimator, for example, it makes a great difference whether the filter comes before or after the decimator. The precise relationships are depicted in Fig. 3.5. The equivalences denoted in Fig. 3.5 have been referred to as the "noble identities" [1, 5]. We note that a common interpretation for both identities is that filtering with $H(z)$ at a given sampling frequency is equivalent to filtering with $H(z^M)$ at M times that sampling frequency, since the impulse response corresponding to $H(z^M)$ is $h(n)$ interpolated by a factor of M.

To finish this section, we stress again the linearity of decimators and interpolators. Thus, any memoryless operation, such as addition and multiplication by a constant, will commute with a decimator or an interpolator. This property is illustrated in Fig. 3.6. In fact, we see that nonlinear memoryless operators would also commute with decimators and interpolators.

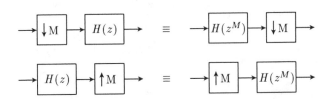

Figure 3.5. Equivalent filtering operations before and after decimation (top) and interpolation (bottom).

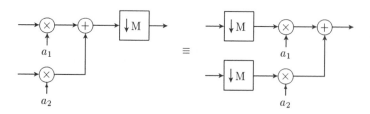

Figure 3.6. Decimators connected before or after memoryless operators are equivalent. The same is true for interpolators.

3.1.3   Polyphase Decompositions

In the analysis of multirate signal processing systems we always come into situations where filter responses are better described in terms of their polyphase components [1]. Let us consider a sequence $h(n)$, with z-transform

$$H(z) = \sum_{n=-\infty}^{\infty} h(n)\, z^{-n}$$

We can always partition the index n into M phases, where each phase is characterized by picking all indices which are identical modulo M. Thus, the kth phase of $h(n)$ is defined by

$$p_k(m) \equiv h(mM + k)$$

We see that $p_k(m)$ can be obtained by passing $h(n)$ through a delay z^{-k} and a $M\!:\!1$ decimator. The original sequence can always be recovered from its polyphase components by

$$h(n) = p_{n \bmod M}(\lfloor n/M \rfloor)$$

where $\lfloor x \rfloor$ means the largest integer not greater than x.

Calling $P_k(z)$ the transform of the kth polyphase component, it is easy to see that [1]

$$H(z) \;=\; \sum_{k=0}^{M-1} z^{-k} P_k(z^M)$$

$$= \sum_{k=0}^{M-1} z^{-k} \sum_{l=-\infty}^{\infty} h(lM + k)z^{-lM}$$

For example, assume that $h(n)$ is FIR, being nonzero from $n = 0$ to $n = 8$, and that we want a three-phase decomposition, i.e., $M = 3$. From the above equations, we see that the decomposition would be

$$
\begin{aligned}
H(z) &= [h(0) + h(3)z^{-3} + h(6)z^{-6}] \\
&\quad + z^{-1}[h(1) + h(4)z^{-3} + h(7)z^{-6}] \\
&\quad + z^{-2}[h(2) + h(5)z^{-3} + h(8)z^{-6}] \\
&= P_0(z^3) + z^{-1}P_1(z^3) + z^{-2}P_2(z^3)
\end{aligned}
$$

3.2 Filter Banks

The fundamental parts of a subband signal processing systems are the analysis and synthesis filter banks, as we have discussed in the previous section. The analysis filter bank should do a good job of separating the incoming signal into its constituent subband signals, and the synthesis filter bank should be able to recover a good approximation to the input signal from the subband signals.

3.2.1 Structures FB-I and FB-II

The basic block diagram of a subband processing system is shown in Fig. 3.7. The analysis filter bank is composed of filters $H_0(z)$ to $H_{M-1}(z)$, and the synthesis filter bank is formed by filters $F_0(z)$ to $F_{M-1}(z)$. The reconstructed signal $\hat{x}(n)$ is obtained simply by adding all the outputs of the synthesis filters. The subband signals $X_0(m)$ to $X_{M-1}(m)$ are generated by decimating the outputs of the analysis filters by a factor of M. The processed subband signals $\hat{X}_0(m)$ to $\hat{X}_{M-1}(m)$ must be interpolated by the same factor, so that we return to the original sampling rate in the reconstructed signal $\hat{x}(n)$.

As we have mentioned in the previous section, the reason for subsampling the subband signals is to avoid a growth in the total number of samples to be processed. Because each subband signal has $1/M$th the original sampling rate and there are M subbands, the total rate in the set of all subband signals is identical to the input and output sampling rates. These filter banks, in which the decimation factor is

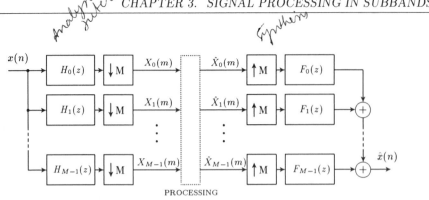

Figure 3.7. Typical subband processing system, using the filter banks in the form FB-I.

equal to the number of subbands, are usually referred to as critically-subsampled or maximally-decimated filter banks [1, 5]. In all applications discussed in this book, either critical decimation is employed, or we have single-rate filter banks with no decimators.

Another restriction that is implicitly imposed in Fig. 3.7 is that all subband filters have their bandwidths close to π/M. There are practical applications in which we do not want all subbands to have equal bandwidths. In such cases, we would have to decimate each subband by a different factor, in order to minimize the aliasing effects. In general, if the kth subband had width π/L, it could be decimated by a factor of L, and the sum of all bandwidths should be approximately equal to π. In this chapter, we will only consider filter banks with subbands of equal widths; they are usually referred to as uniform filter banks. In Chapter 6 we will discuss nonuniform filter banks and their applications.

In Fig. 3.7 the subband signals are all simultaneously decimated, that is, when $n = mM$ the outputs of all analysis filters are $X_k(m)$, for $k = 0, 1, \ldots, M - 1$. We could also subsample the subbands sequentially. When $n = mM$ we would have a sample of the subband number $M - 1$ into $X_{M-1}(m)$. At the next instant $n = nM + 1$, we would have a sample of subband number $M - 2$ into $X_{M-2}(m)$, and so on, until $n = mM + (M - 1)$, when we would get a sample of subband number 0 into $X_0(m)$. At the next sample, the whole process would be repeated again. This can be implemented, in terms of a block diagram, by preceding the decimators with appropriate delay units, as shown in Fig. 3.8. To compensate for the delays

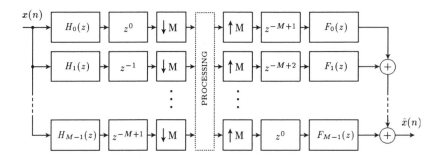

Figure 3.8. Typical subband processing system, with sequential down-sampling and up-sampling. This is the form FB-II.

introduced before the decimators, we would need delays after the interpolators, as shown in Fig. 3.8. We will refer to this structure as the FB-II filter bank.

3.2.2 Signal Reconstruction

The fundamental property that the subband processing system of Fig. 3.7 should have is good signal reconstruction, that is, if the subband signals are not modified, i.e., if $\hat{X}_k(m) = X_k(m)$, $k = 0, 1, \ldots, M - 1$, then the output signal $\hat{x}(n)$ should approximate $x(n - D)$, where D is an acceptable processing delay that depends on the application. In this way, the only modifications in the reconstructed signal would be due to the processing of the subband signals.

In order to obtain the necessary condition on the filter responses for signal reconstruction, let us start by noting that the output of the kth analysis filter is given by [1]

$$X_k(m) = \sum_{n=-\infty}^{\infty} x(n)\, h_k(mM - n) \qquad (3.7)$$

where $h_k(n)$ is the impulse response of the kth analysis filter. The reconstructed signal can be written as a function of the processed subband signals as

$$\hat{x}(n) = \sum_{k=0}^{M-1} \sum_{m=-\infty}^{\infty} \hat{X}_k(m)\, f_k(n - mM) \qquad (3.8)$$

where $f_k(n)$ is the impulse response of the kth synthesis filter.

When we do not modify the subband signals, we have

$$\hat{X}_k(m) = X_k(m), \; \forall k$$

Then (3.7) can be substituted into (3.8), with the result [6]

$$\hat{x}(n) = \sum_{l=-\infty}^{\infty} x(l)\, h_T(n,l) \tag{3.9}$$

where the time-varying impulse response of the total system is given by

$$h_T(n,l) \equiv \sum_{k=0}^{M-1} \sum_{m=-\infty}^{\infty} f_k(n-mM)\, h_k(mM-l) \tag{3.10}$$

We can obtain perfect reconstruction (PR) if and only if $h_T(n,l) = \delta(n-l-D)$, that is,

$$\sum_{k=0}^{M-1} \sum_{m=-\infty}^{\infty} f_k(n-mM)\, h_k(mM-l) = \delta(n-l-D) \tag{3.11}$$

where D is just a delay, which must be included if we want the analysis and synthesis filters to be causal. If (3.11) holds, it follows that

$$\hat{x}(n) = x(n-D)$$

Although (3.11) seems simple, finding a set of realizable filters $\{h_k(n)\}$ and $\{f_k(n)\}$ that satisfy that condition is far from being an easy task. In fact, less than a decade has passed since the first realizable exact solutions to (3.11) were obtained. When we look at the FB-II structure, though, we see that it allows for a trivial PR solution, in which all analysis and synthesis filters are replaced by unity gains. Then, the subband signals $X_k(m)$, $k = 0, 1, \ldots, M-1$ would be just the M polyphase components [1] of the input signal, and the synthesis filter would just append them to reconstruct exactly the original signal, with a delay of $M-1$ samples. This idea is depicted in Fig. 3.9 for $M = 3$.

Such a trivial solution is of no practical value, because the subband signals would have the same bandwidth as the original signal. However, it is quite important theoretically because it shows us two fundamental properties of filter banks: first, even

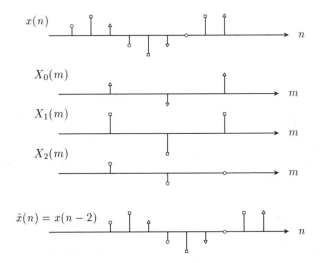

Figure 3.9. Perfect reconstruction with no filtering in the FB-II structure, for $M = 3$.

very large amounts of aliasing generated in the analysis filter bank can be totally cancelled in the synthesis filter bank; second, filters allowing perfect reconstruction can have short FIR responses. We shall see in Section 3.4 that the PR property is not a privilege of trivial and useless solutions. In fact, good FIR filter banks with perfect reconstruction can be designed and efficiently implemented.

3.2.3   Computational Complexity

When we run the analysis filter bank in Fig. 3.7 and Fig. 3.8, the decimators imply that we only need to compute one out of every M output samples, and this reduces the computational complexity by a factor of M, if the filters are FIR. However, because there are M filters, we conclude that running all the filters in the analysis filter bank has the same complexity as running a single filter without decimation. Thus, if the impulse responses of the analysis filters have length L, we need L multiplications and $L - 1$ additions to run the whole analysis filter bank for each new incoming sample, in a direct form implementation. The synthesis filter bank has the same complexity because the input to each synthesis filter comes from a $1:M$ interpolator; for every M samples, only one is nonzero.

Thus, we see that in the analysis/synthesis filter bank with FIR filters of length L, the total number of operations necessary to run it is approximately $4L$ per input sample, if the filters are implemented in direct form. This is quite within the reach of modern digital signal processing chips for the sampling rates involved in speech coders, speech scramblers, and voiceband modems, for example, for a small number of bands. For large M, though, the computational complexity may be prohibitive. This is because if M is increased, then L will have to grow proportionally, since the bandwidth of each filter must be π/M.

We recall from Chapter 2 that for block transforms the complexity is on the order of $\log_2 M$ operations per input sample. This seems to indicate that subband coding would be much more computationally intensive than transform coding for the same number of bands. In the following sections we will see that there are filter bank structures in which a block transform can be inserted as part of the filter bank. This can make the computational complexity of filter banks proportional to $\log_2 M$ instead of proportional to M. In fact, as we will see in Chapters 4 and 5, lapped transforms are actually filter banks with fast computational algorithms, and so they allow the efficient implementation of subband processing systems with a large number of bands.

3.2.4   DFT Filter Banks

Returning to the FB-I structure of Fig. 3.7, it is easy to verify that if $H_k(z)$ and $F_k(z)$ are ideal band-pass filters, then (3.11) will be satisfied. Therefore, an approximation to perfect reconstruction could be obtained, in principle, simply by designing good band-pass filters with any of the standard filter design techniques. For example, we could use the window approach or the Parks-McClellan method to design good FIR band-pass filters [7].

The advantages of using FIR filters is that they can easily be designed to have linear phase, i.e., a constant group delay. If we force all channels to have the same delay (that is, $F_k(z)H_k(z)$ has a group delay D for all k), then the analysis/synthesis systems in the FB-I or FB-II structures will have no phase distortion, and so the uncanceled aliasing will lead only to amplitude distortions [1]. This is why FIR filters are much more frequently used in practice. If the signal to be processed has finite length, though, as is the case in image processing, PR systems with IIR

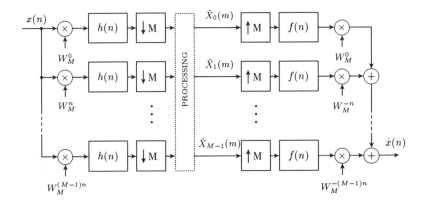

Figure 3.10. DFT filter bank with complex modulators.

filters can be designed, using a combination of causal and anticausal processing [8]. Throughout this book, we shall consider only FIR filter banks.

If we design each filter in the analysis and synthesis filter bank independently, then the computational complexity will be proportional to the number of bands, as discussed in the previous section. The first step toward more efficient implementations of the filter bank was the development of the DFT filter bank [1, Chapter 7], which is depicted in Fig. 3.10.

In the DFT filter bank, each subband first passes through a complex modulator to shift its center frequency to the origin. Then, a low-pass filter $h(n)$ is used before decimation. On the synthesis side, a low-pass filter $f(n)$ is used after interpolation, and the filtered subband is brought back from baseband to its original center frequency by means of complex modulation again. Note that the subband signals now are complex, but for a real input $x(n)$ and a real low-pass filter $h(n)$ it is easy to see that

$$X_{M-k}(m) = X_k^*(m)$$

and so the total number of independent values in the real and imaginary parts of $\{X_0 \cdots X_{M-1}\}$ is M. Therefore, there is no information growth.

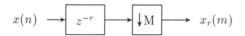

Figure 3.11. Generation of the auxiliary sequence $x_r(m)$.

The kth subband signal in the DFT filter bank of Fig. 3.10 can be written as [1]

$$X_k(m) = \sum_{n=-\infty}^{\infty} h(mM - n)\, x(n)\, W_M^{kn} \qquad (3.12)$$

with $W_M \equiv e^{-j2\pi/M}$. Note that in [1] it was used $W_M = e^{j2\pi/M}$, but this leads just to a change of sign in the imaginary part of $X_k(m)$. If we decompose the index n in the form $n = lM - r$, the above equation is equivalent to

$$X_k(m) = \sum_{r=0}^{M-1} u_r(m)\, W_M^{kr} \qquad (3.13)$$

where

$$u_r(m) \equiv \sum_{l=-\infty}^{\infty} h(mM - lM + r)\, x(lM - r) \qquad (3.14)$$

Thus, for every subband clock instant m, we see from (3.13) all filter bank outputs can be computed by a length-M FFT applied to the block $[u_0(m) \cdots u_{M-1}(m)]^T$. Furthermore, let us consider the polyphase decomposition of the low-pass filter $h(n)$

$$p_i(m) = h(mM + i), \quad i = 0, 1, \ldots, M - 1$$

and let us also define a sequence $x_r(m)$ according to Fig. 3.11, that is

$$x_r(m) \equiv x(mM - r), \quad r = 0, 1, \ldots, M - 1$$

Then, (3.14) can be written in the form

$$u_r(m) = \sum_{l=-\infty}^{\infty} p_r(m - l)\, x_r(m) \qquad (3.15)$$

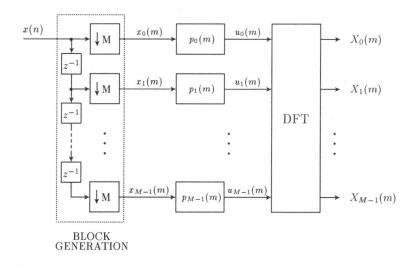

Figure 3.12. DFT filter bank with input blocking and FFT, analysis bank.

We see, therefore, that the sequence to be transformed $\{u_r(m)\}$, can be obtained by filtering $x_r(m)$ with the filter $p_r(m)$ at the subband sampling rate, i.e., $1/M$th the original sampling rate. Note that if the filter $h(n)$ has length $L = NM$, then each of its polyphase components $p_r(m)$ has length N.

Using the results in (3.12) to (3.15), we can build the block diagram of the analysis section of the DFT filter bank as in Fig. 3.12. The synthesis filter bank can be implemented by means of the polyphase decomposition of the synthesis filter $f(n)$, as shown in Fig. 3.13. The polyphase components of $f(n)$ are

$$q_i(m) \equiv f(mM + i), \quad i = 0, 1, \ldots, M - 1$$

The proof that the synthesis filter bank can be put in the form shown in Fig. 3.13 is quite similar to the developments in (3.12) to (3.15) [1, 2, 9]. Note that in some references the analysis filter uses an inverse DFT and the synthesis filter uses a direct DFT [2, 5, 10] because a modulation by W_M^{kn} was used. As we have already mentioned, this difference is immaterial because it only affects the signs of the imaginary parts of the subband signals.

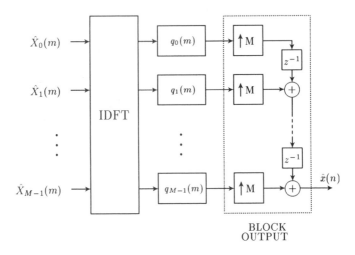

Figure 3.13. DFT filter bank with input blocking and FFT, synthesis bank.

There are several interesting properties of the DFT filter bank of Fig. 3.12 and Fig. 3.13. One is in the part labeled "block generation." Because its outputs are decimated by a factor of M, the next sample from the outputs of the decimators will be available when M new input samples from $x(n)$ are shifted into the memory formed by the $M-1$ delay elements. Therefore, the input signal is actually being processed in blocks of M samples, just as in transform coding, for example. Also, the part of the figure labeled "block output" shows that when $n = mM$ we have a new block available from the outputs of the interpolators. This block is added to the M-sample buffer formed by the $M-1$ delay elements, and for the next $M-1$ samples the buffer is sequentially shifted to the output $\hat{x}(n)$. Therefore, the blocks labeled "block generation" and "block output" represent a multirate implementation of a block processing system.

Another important point to be noted from the DFT filter bank is that the filters $p_r(m)$ and $q_r(m)$ operate at the subband sampling rate (which we could also call the block rate). Thus, the input to $p_r(m)$, for example, is formed by the sequence of the rth samples of consecutive input blocks. A one-sample memory in $P_r(z)$ means a memory from one block to the next. Coupling this with a SRFFT implementation of the DFT, for example, leads to a much lower computational complexity than the general FB-I structure of Fig. 3.7. For example, with 32 channels and filters of length 128, running the analysis/synthesis filter banks with the FB-I structure would require about 512 arithmetic operations per input sample. With the DFT filter bank based on $h(n)$ and $f(n)$ with lengths equal to 128 (and hence, with the potential for the same filtering performance, in terms of passband and stopband gains), the number of operations per input sample would drop to about 24, a reduction of more than an order of magnitude.

The DFT filter bank in Fig. 3.12 and Fig. 3.13 shows us also immediately that in order to obtain perfect reconstruction it is sufficient (but not necessary) that $P_r(z) Q_r(z) = z^{-D}$. Thus, the filter $Q_r(z)$ should be the inverse of the filter $P_r(z)$, within a delay. For example, if $P_r(z) = Q_r(z) = 1$, we will have perfect reconstruction in the DFT filter bank, with the analysis and synthesis filters given by

$$f(n) = h(n) = \begin{cases} 1, & \text{if } 0 \le n \le M - 1 \\ 0, & \text{otherwise} \end{cases}$$

Note that in this case the DFT filter bank becomes a block transform processor using the DFT. We will return to this issue in Section 3.5. The equation above shows us that there is a set of FIR filters that will lead to perfect reconstruction in a DFT filter bank. Unfortunately, *all* possible FIR solutions for $M > 2$ are those in which the polyphase components of $h(n)$ and $f(n)$ have order zero [5], that is, $h(n)$ is a length-M window and $f(n)$ is the corresponding inverse window. In this way, $h(n)$ and $f(n)$ will not be good low-pass filters, and so perfect reconstruction with the DFT filter bank is not feasible in practice.

If we restrict the low-pass filters $h(n)$ and $f(n)$ to be FIR, both having lengths equal to $L \gg M$, it may be possible to keep the aliasing error to within tolerable limits. In fact, using L on the order of $10M$ and a standard window design for $h(n)$, it is possible to keep the reconstruction error below 0.1% [1, Section 7.3]. However, using the aliasing cancellation techniques described in the next section, it is possible to achieve similar performances in terms of amplitude distortion and subband separation, but without aliasing. Furthermore, with lapped transforms, it is possible to design filter banks with good subband separation and perfect reconstruction, and still keeping the fast computation property of DFT filter banks. Thus, we will not consider further the design of $h(n)$ and $f(n)$ for DFT filter banks, but the structure of Fig. 3.12 will be essential for the developments of the following sections and those of Chapters 4 and 5.

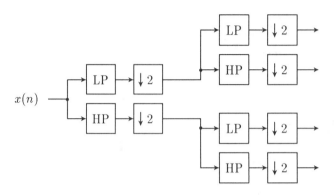

Figure 3.14. Tree connection of two-band filter banks generating a 2^s-band filter bank. LP and HP mean low-pass and high-pass filters, respectively.

3.3 Quadrature Mirror Filters

In the search for perfect aliasing cancellation in the analysis/synthesis filters of Fig. 3.7, the initial studies were concentrated for the two-channel bank, i.e., $M = 2$. This is not very restrictive because we can generate a filter bank with $M = 2^s$ channels by a tree of two-channel banks. First, we split the full band $[0, \pi]$ into two, then each one is split in two again, and so on. With s levels in such a tree, we would have 2^s subbands at the final level, as shown in Fig. 3.14. This arrangement may lead to some problems, though, as we will see in the end of this section.

3.3.1 Two-channel QMF Banks

The analysis of the aliasing effects for a two-channel filter bank is relatively simple when $M = 2$. Using the z-domain relationships for the decimators and interpolators in (3.2) and (3.4), we can obtain a simple expression for the z-transform of the output signal in Fig. 3.7 [1, 11]

$$
\begin{aligned}
\hat{X}(z) &= \frac{1}{2} \left[H_0(z)\, F_0(z) + H_1(z)\, F_1(z) \right] X(z) \\
&\quad + \frac{1}{2} \left[H_0(-z)\, F_0(z) + H_1(-z)\, F_1(z) \right] X(-z)
\end{aligned}
\tag{3.16}
$$

The aliasing component $X(-z)$ can be cancelled in the above equation with the following choice for the synthesis filters, first suggested in [12]

$$H_1(z) = H_0(-z), \; F_0(z) = H_0(z), \; F_1(z) = -H_0(-z) \qquad (3.17)$$

Note that the above equation implies $H_1(e^{j\omega}) = H_0(e^{j(\omega+\pi)})$. Thus, if H_0 is a low-pass filter, then H_1 is a high-pass filter with a symmetrical frequency response, since $\omega \to \omega + \pi$ is a low-pass to high-pass transformation.

When the analysis and synthesis filters satisfy (3.17), they are referred to as *quadrature mirror filters* [12] (QMF), due to their mirror-image frequency responses with respect to the frequency $\omega = \pi/2$.

With the restrictions in (3.17), (3.16) simplifies to

$$\hat{X}(z) = \frac{1}{2} \left[H_0^2(z) - H_1^2(z) \right] X(z)$$

from which we see that the aliasing term has been completely cancelled. Ideally, we would like $\hat{x}(n)$ to be a delayed copy of $x(n)$. For this to occur, we should have

$$H_0^2(z) - H_0^2(-z) = 2z^{-D} \qquad (3.18)$$

where D is some acceptable delay.

If the conditions in (3.17) and (3.18) were satisfied, we would have not only total aliasing cancellation but also perfect signal reconstruction. Unfortunately, there are no filters $H_0(z)$ and $H_1(z)$ with finite order greater than one that can satisfy both conditions [1, 11]. Perfect reconstruction is only possible with trivial solutions such as $H_0(z) = \sqrt{1/2}\,(1 + z^{-1})$ or with ideal filters, none of which are suitable in practice. However, it is possible to find a FIR filter $H_0(z)$ such that when we impose (3.17) to avoid aliasing, (3.18) is almost exactly satisfied, as we shall see later.

An important consequence of the QMF conditions in (3.17) is that they generate an efficient polyphase implementation [1]. In order to see that, let us bring (3.17) to the time domain:

$$h_1(n) = (-1)^n h_0(n), \; f_0(n) = h_0(n), \; f_1(n) = -(-1)^n h_0(n) \qquad (3.19)$$

When we write $H_0(z)$ in terms of its polyphase components

$$H_0(z) = P_0(z^2) + z^{-1} P_1(z^2)$$

INPUT BLOCK

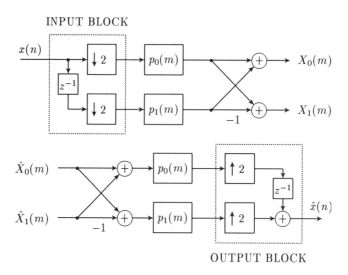

OUTPUT BLOCK

Figure 3.15. Efficient polyphase implementation of QMF filter banks. Top: analysis; bottom: synthesis.

where

$$p_0(m) = h_0(2m)$$
$$p_1(m) = h_0(2m + 1)$$

it becomes clear that the even phase components of $h_0(n)$ and $h_1(n)$ are equal, whereas their odd phase components are related simply by a negative sign. A similar relationship holds for the synthesis filters.

Using the above relationships, the QMF filter bank can be implemented by the structure shown in Fig. 3.15. Comparing the polyphase QMF structure with that of the DFT filter bank in Fig. 3.12 and Fig. 3.13, and recalling that the DFT matrix of order two is

$$\begin{bmatrix} 1 & 1 \\ 1 & -1 \end{bmatrix}$$

we see that the QMF filter bank is actually a special case of the DFT filter bank, with identical polyphase filters for analysis and synthesis, and $M = 2$.

In the QMF filter bank, only the filter $H_0(z)$ can be freely chosen; the others must satisfy (3.17). If we assume a linear phase FIR filter, for example, we have

$$h_0(L - 1 - n) = h_0(n)$$

where L is the filter length. Then, the QMF flatness condition in (3.18) can be written in the Fourier domain as

$$|H_0(e^{j\omega})|^2 + |H_0(e^{j(\omega+\pi)})|^2 = 2 \qquad (3.20)$$

where we have assumed that L is even. If L is odd, the overall gain at $\omega = \pi/2$ is forced by (3.17) to be zero [1], and so useful QMF filters have an even number of taps.

The design problem for a FIR QMF bank involves two approximation problems: first, $H_0(e^{j\omega})$ should be close to the ideal filter with a cutoff frequency of $\pi/2$; second, (3.20) must be approximately satisfied. If we use a standard FIR filter design technique, such as the window method, good results can be obtained. By adjusting the cutoff frequency ω_c of the ideal filter that will be windowed, and also the window parameters, it is possible to approximate (3.20) closely. In Fig. 3.16 we have an example of this idea. We have used a Kaiser window [7] with $\beta = 3.2$ and length $L = 32$, and we have chosen $\omega_c = 0.5192\pi$. We note that the transition bandwidth is about 0.09π and the stopband attenuation is -38 dB. The amplitude reconstruction error, in terms of deviation from (3.20), is plotted in Fig. 3.17. We see that the total ripple is ± 0.16 dB, which would be quite acceptable in many applications.

It is possible to reduce the ripple in the total power response by defining an error function of the form

$$E = \int_{\omega_s}^{\pi} |H_0(e^{j\omega})|^2 d\omega + \alpha \int_0^{\pi} [|H_0(e^{j\omega})|^2 + |H_0(e^{j(\omega+\pi)})|^2 - 2]^2 d\omega \qquad (3.21)$$

where the first term is the energy in the stopband $[\omega_s, \pi]$, and the second term is a measure of the reconstruction ripple. A nonlinear optimization algorithm could be used to search the space of coefficients $h_0(n)$ in order to minimize E. This has been carried out by Johnston in [13]. As an example, consider the filter 32D [1, Appendix 7.1], whose magnitude response is plotted in Fig. 3.18. The characteristics of the 32D filter are quite similar to those of our Kaiser window filter, but the

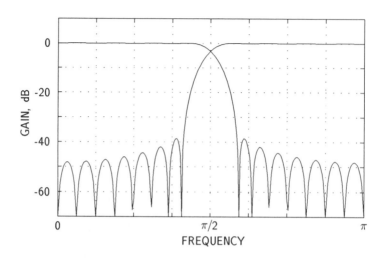

Figure 3.16. Magnitude responses of the QMF filters designed with a Kaiser window.

reconstruction ripple of the 32D filter is much lower, as shown in Fig. 3.19 (around ±0.015 dB).

Another technique for the minimization of the error in (3.21) was presented in [14], where the reconstruction ripple is written in the time-domain, and the analysis and synthesis are iteratively optimized. This approach is more stable than the one in [13] because a major part of the optimization task is put in terms of an eigenvector-eigenvalue problem.

One word of caution is necessary before we leave the two-channel QMF filter bank. Recall that one of the arguments why a two-channel QMF bank may be useful is because we can build a tree of QMF filter banks, as in Fig. 3.14, and so effectively split the input signal into more than two subbands. In this way, the aliasing cancellation of the QMF structure would be preserved, and if we started with QMF filters having a small reconstruction ripple, the overall tree structure would also have a small ripple (at most, the ripple in dB would be multiplied by the number of levels in the tree). However, aliasing cancellation would only occur at the synthesis filters, and the cascade of several decimators and nonideal filters in the analysis tree structure of Fig. 3.14 may lead to some form of aliasing build-up, which would degrade the frequency responses of the filters.

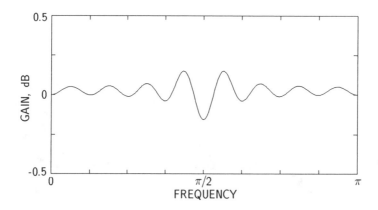

Figure 3.17. Reconstruction error with the Kaiser window QMF filters.

For example, consider an eight-band filter bank implemented as a tree of three two-band QMF stages. All stages are built with 16B filters [1, 13]. The magnitude frequency responses of some of the equivalent subband filters are shown in Fig. 3.20. We can see that the low-pass and high-pass filters in the tree turn out to be good filters, with a stopband attenuation of about 45 dB. However, in subband number 5 the filter shows a spurious sidelobe at only −18 dB.

Although the aliasing that would be allowed by spurious sidelobes would eventually be cancelled in the synthesis filter, we must recall that in most applications we process the subbands, so that perfect aliasing cancellation is precluded. Thus, it is important that the aliasing level be small to begin with. Furthermore, in subband coding, for example, the quantization noise that will be present in the received subband signals will be emphasized by responses with spurious sidelobes such as those in Fig. 3.20. The same problem occurred in [9] when it was tried to force quasi-perfect reconstruction in a DFT filter bank.

There are applications where the number of bands may be greater than eight, as in speech coding and speech scrambling. For those, a filter bank composed of a tree of two-channel QMF filters can only be used if the original filters are long ($L > 32$), in order to avoid the spurious sidelobes present in Fig. 3.20. A good alternative would be if QMF filter banks could be designed for $M > 2$. This is the subject of the next subsection.

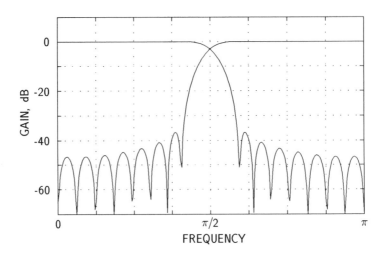

Figure 3.18. Magnitude responses of the Johnston 32D QMF filter bank.

3.3.2   QMF Banks for $M > 2$

The next logical step after the development of the two-channel QMF filter bank would be to try to generalize the QMF properties for $M > 2$. Another motivation is that, as we saw in the previous subsection, tree structures based on two-channel QMF filters can only be used if the filter impulse responses are long, and this would lead to high computational complexities. Thus, a parallel M-band structure as in Fig. 3.7, with filters designed for total aliasing cancellation, might be more efficient than tree structures. Furthermore, if the band-pass filters $H_0(e^{j\omega})$ to $H_{M-1}(e^{j\omega})$ are generated from modulated versions of a single low-pass prototype filter, as was the case with the DFT filter bank, we might expect fast transform-based implementations.

The first proposal for a QMF-like filter bank for $M > 2$ was presented independently in [15], with the denomination "pseudo-QMF filter bank," and in [16], where they were referred to as "polyphase quadrature filters." The fundamental idea is that the filters are designed so that only the aliasing from adjacent bands is cancelled. This makes sense, because the transition bandwidth of each one of the subband filters is unlikely to extend to a frequency range beyond that of the neighboring filters. Then, the uncanceled aliasing would be at the same level as

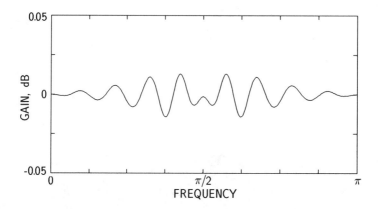

Figure 3.19. Reconstruction error with the 32D QMF filters.

the stopband attenuation of the subband filters, which can be made low enough to be negligible. Furthermore, it was suggested in [15] that the band-pass filters be modulated versions of a single low-pass filter, for easier implementation [17].

Other alternatives for M-band filter banks with nearly perfect aliasing cancellation and FIR analysis and synthesis filters were presented [18, 19]. They can all be put in a common form [20], where the synthesis filters are defined by

$$f_k(n) = h(n)\cos\left[\left(k + \frac{1}{2}\right)\left(n - \frac{L-1}{2}\right)\frac{\pi}{M} + \phi_k\right] \tag{3.22}$$

for $k = 0, 1, \ldots, M - 1$, and $n = 0, 1, \ldots, L - 1$, where M is the number of bands, L is the length of the filters, and the parameters ϕ_k control the relative phases of the modulating cosines. The analysis filters are obtained by time-reversing the synthesis filters, in the form

$$h_k(n) = f_k(L - 1 - n) \tag{3.23}$$

The filter $h(n)$ that is used to derive all filters is commonly referred to as the low-pass prototype.

The form of the modulating functions in (3.22) is easy to understand intuitively. First, the term $(k + 1/2)$ means that the center frequencies of the modulating functions are precisely the center of the intervals $[k\pi/M, (k + 1)\,\pi/M]$. This choice

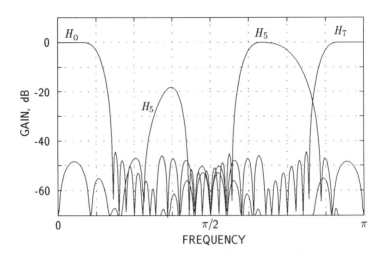

Figure 3.20. Frequency responses of the equivalent filters for subbands numbers 0, 5, and 7, for a tree-structured QMF bank composed of fourteen 16B filters. Note the undesirable side lobe in the response of the equivalent filter for subband number 5.

of modulating frequencies was first used in SSB filter banks [1] and is usually referred to as odd channel stacking (an even channel stacking would start, as in the DFT bank, at zero frequency). The term $(L - 1)/2$ that is subtracted from n is just a phase reference, with the property that if ϕ_k is a multiple of $\pi/2$ than the filters will have linear phase. For other choices of the phases ϕ_k, the filters will not be linear phase, but the time-reversing property in (3.23) means that the cascade of the analysis and synthesis filter for the kth channel is a linear phase filter with length $2L - 1$. Therefore, the whole filter bank is linear phase, with a total delay of $L - 1$ samples from $x(n)$ to $\hat{x}(n)$, and the uncanceled aliasing will only result in amplitude distortions.

In order to achieve aliasing cancellation from neighboring subbands, the phases must be restricted by

$$\phi_{k+1} - \phi_k = (2r + 1)\frac{\pi}{2} \tag{3.24}$$

where r is an integer, and the low-pass prototype must satisfy as closely as possible the "generalized QMF conditions" [15, 16]

$$|H(e^{j\omega})|^2 + |H(e^{j(\pi/M)-\omega})|^2 = 2, \quad 0 < |\omega| < \pi/2M,$$
$$|H(e^{j\omega})|^2 = 0, \quad |\omega| > \pi/M \qquad (3.25)$$

In [16] the choice for the angles was

$$\phi_k = (-1)^k \frac{\pi}{4} \qquad (3.26)$$

In fact, this choice is quite important because it may lead to perfect reconstruction (PR) filter banks, a fact that was not recognized in [16–20]. We return to this possibility of perfect reconstruction in the next section. Another possible choice is [16, 19]

$$\phi_k = [1 + (-1)^k] \frac{\pi}{4} \qquad (3.27)$$

but this will lead to an overall gain of zero at $\omega = 0$ and $\omega = \pi$. There are applications such as speech coding and scrambling, for example, that these zero responses at dc and half the sampling frequency is not a problem [16, 20]. These zero responses can be avoided by designing $f_0(n)$ and $f_{M-1}(n)$ independently, i.e., by not using (3.22) for the first and last filters in the bank. An example of this approach for $L = 2M$ is the pseudolapped orthogonal transform [21], which has a simple implementation using the DCT.

It is important to note that all known modulated filter banks follow the two basic choices for the angles in (3.26) or (3.27), except for trivial sign changes (like adding π to some of the ϕ_k's). The design of pseudo-QMF filter banks based on (3.22) can be done by small modifications on traditional QMF filter design procedures such as in [13, 14], since the design restriction in (3.25) is quite similar to that in (3.20). However, we see in the next section that it is actually possible to design the low-pass prototype to achieve perfect reconstruction, i.e., $\hat{x}(n) = x(n - L + 1)$. Thus, we shall not discuss in detail the design of pseudo-QMF filter banks.

3.4 Perfect Reconstruction

We say that a filter bank has the perfect-reconstruction property (PR) if its time-varying impulse response in (3.10) is an impulse, i.e., if $h_T(n, l) = \delta(n - l - D)$. In such a system, not only aliasing is totally cancelled, but there is no amplitude or group delay distortions. As we saw in the previous sections, these distortions can be made small enough with QMF or pseudo-QMF filter banks. Nevertheless, we would certainly prefer to use PR FIR filter banks if they could be designed within the same computational complexity of the QMF or pseudo-QMF solutions. This is indeed the case, and the purpose of this section is to review the design techniques for PR filter banks.

3.4.1 Two-channel PR Filter Banks

There is a basic questions that arises when we look at the two-channel QMF conditions in (3.17): is that choice of filters unique, that is, is there any other choice that will also allow aliasing cancellation? Fortunately, the answer to this question is yes, and a new set of relationships was reported independently in [11] and [22]. After [11], we will refer to these filters as *conjugate quadrature filters* (CQF), which are defined by the following choices for the analysis and synthesis filters

$$H_1(z) = z^{-(L-1)} H_0(-z^{-1}), \quad F_0(z) = H_1(-z), \quad F_1(z) = H_0(-z) \qquad (3.28)$$

where L is the length of the filters (actually, we have changed the signs of both $H_1(z)$ and $F_1(z)$, compared to [11], but this is irrelevant because (3.16) remains valid). As in the QMF case, we assume that L is even Note that, in the time domain, the restrictions in (3.28) correspond to

$$
\begin{aligned}
h_1(n) &= -(-1)^n h_0(L-1-n) \\
f_0(n) &= h_0(L-1-n) \\
f_1(n) &= (-1)^n h_0(n) = h_1(L-1-n)
\end{aligned}
\qquad (3.29)
$$

With the CQF constraint in (3.28), (3.16) becomes

$$\hat{X}(z) = \frac{1}{2} \left[H_0(z) H_0(z^{-1}) + H_0(-z) H_0(-z^{-1}) \right] X(z) \qquad (3.30)$$

Now, let us assume that $H_0(z)$ is a spectral factor of a half-band filter $U(z)$. Recall that a half-band filter is a linear phase filter that satisfies $u(0) = 1$ and $u(2l) = 0$, $l = \pm 1, \pm 2, \ldots$ [1]. Then,

$$H_0(z) H_0(z^{-1}) = U(z) \qquad (3.31)$$

Because $U(z)$ is a half-band filter, it can be written in the form

$$U(z) = 1 + z^{-1}P(z^2) \qquad\qquad (3.32)$$

from which it is clear that its first polyphase component is just a constant in the frequency domain (because it is an impulse in the time domain). Substituting (3.31) and (3.32) in (3.30), we obtain

$$\hat{X}(z) = \frac{1}{2}\left[U(z) + U(-z)\right] = -z^{-(L-1)}X(z)$$

Thus, the reconstructed signal is an exact replica of the input signal, with a delay of $L - 1$ samples.

We see then that the design of a PR FIR filter bank is relatively simple. For a given length L, we should design a filter $u(n)$ of length $2L - 1$ that is a half-band filter. The traditional windowing technique, for example, will guarantee that. Then, we should use a spectral factorization technique to find $H_0(z)$. The problem is that not all half-band filters can be factored as in (3.31). It is easy to see from (3.31), though, that the factorization will be possible if all zeros of $U(z)$ occur in reciprocal pairs and if all zeros on the unit circle are double. This is equivalent to $U(e^{j\omega}) \geq 0$ for all ω. A simple way to achieve that is to modify $U(z)$ by

$$U(z) \leftarrow U(z) - \delta_s$$

or, in the time domain

$$u(0) \leftarrow u(0) - \delta_s$$

where $\delta_s = \min\{U(e^{j\omega})\}$. If a modified Parks-McClellan program were used to design the half-band filter $U(z)$, than δ_s would be the stopband ripple. With this new modified $U(z)$, we should verify that the stopband attenuation of $U(e^{j\omega})$ is twice the desired attenuation, in decibels, in view of (3.31).

After the above step we compute the zeros of $U(z)$ and write it in the form

$$U(z) = G \prod_{l=1}^{L-1}(z - z_l)\left(z^{-1} - z_l\right)$$

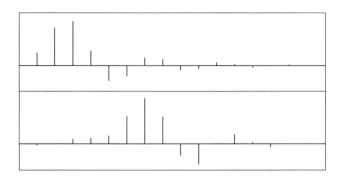

Figure 3.21. Impulse responses of 16-tap CQF filters. Top: minimum phase solution;
bottom: one of the nonminimum phase solutions.

Then, $H_0(z)$ will be given by

$$H_0(z) = \sqrt{G} \prod_{l=1}^{L-1} (z^{-1} - z_l) \tag{3.33}$$

In this last equation, there can be up to 2^{L-1} possible ways of constructing $H_0(z)$, because for each z_l we may choose either z_l or z_l^{-1}. If we choose all poles inside the unit circle, we would have a minimum-phase $H_0(z)$, and that was the suggestion in [22]. In this case, $F_0(z)$ would be the corresponding maximum-phase filter.

If we pick the minimum-phase $H_0(z)$, the larger values of $h_0(n)$ will occur near $n = 0$, and this may lead to blocking effects. Therefore, as suggested in [11], we should pick a mix of zeros z_l inside and outside the unit-circle in (3.33), so as to bring the larger values of $h_0(n)$ close to the center $[n = (L-1)/2]$, with decaying values of $h_0(n)$ as we approach $n = 0$ or $n = L-1$. In this way, the blocking effects will be negligible.

As an example, consider $L = 16$. We have designed $u(n)$ from a length-31 Kaiser window [7] with $\beta = 7.5$, and using $\delta_s = 0.0001$. The minimum-phase solution for $h_0(n)$ is shown in Fig. 3.21, where we see the large values of $h_0(n)$ near $n = 0$. A solution with nonminimum phase is also shown in Fig. 3.21. Note that the values of $h_0(n)$ decay smoothly to zero at its boundaries. Therefore, even with the phase

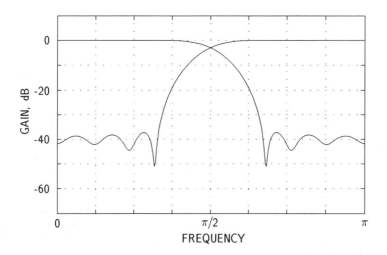

Figure 3.22. Magnitude frequency responses $|H_0(e^{j\omega})|$ and $|H_1(e^{j\omega})|$ for the 16-tap CQF filters of Fig. 3.21.

response of $H_0(z)$ being nonlinear, we should expect better result in practice with the nonminimum phase filter, because it will lead to virtually no blocking effects. The magnitude frequency responses $|H_0(e^{j\omega})|$ and $|H_1(e^{j\omega})|$ (which are the same for both cases) are shown in Fig. 3.22.

We should note that CQF filters are also power-complementary, that is, they satisfy

$$|H_0(e^{j\omega})|^2 + |H_1(e^{j(\omega+\pi)})|^2 = 2 \tag{3.34}$$

Note that now the filters H_0 and H_1 are not linear phase. In fact, power-complementary PR filters for $M = 2$ cannot have both linear phase [23] responses. Note that, since (3.29) would force $h_1(n)$ to be linear phase if $h_0(n)$ were linear phase, we conclude that CQF filters cannot have liner phase. Finally, another important characteristic of CQF filters is that the polyphase implementation of Fig. 3.15 cannot be used. This is because the filters $h_0(n)$ and $h_1(n)$ do not have the same polyphase components. Thus, the computational complexity of CQF filter banks is approximately twice of that of QMF filter banks, for the same filter length.

INPUT BLOCK

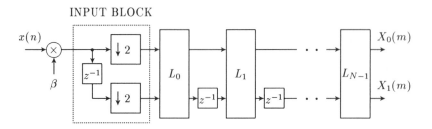

Figure 3.23. Lattice structure for perfect-reconstruction, two-channel filter banks, analysis section.

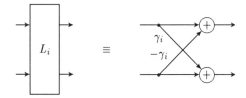

Figure 3.24. Detail of each lattice section L_i in Fig. 3.23. For the synthesis lattices L_i', the signs marked with asterisks should be used.

An efficient implementation of CQF filters is the lattice structure proposed in [23]. These structures are in fact a particular case of canonical structures presented in [24] for arbitrary M, which will be discussed in the next section. The lattice structure is shown in Fig. 3.23, which is a cascade of two kinds of blocks: simple delay elements and memoryless lattice sections.

The structure of each lattice section is shown in Fig. 3.24. When we compare the lattice structure of Fig. 3.23 with the QMF polyphase implementation in Fig. 3.15, it is clear that in the lattice structure there is an extra degree of freedom. In the QMF structure, each of the two channels is independently processed by the polyphase component filters $p_0(n)$ and $p_1(n)$, whereas in the lattice structure there can be cross-coupling among the channels within the lattice.

In order to obtain the filters $H_0(z)$ and $H_1(z)$ of the lattice structure in Fig. 3.23, let us move for a while the $2:1$ decimators to the end of the structure. According to Fig. 3.5, this would change all the delay units but the leftmost to z^{-2}. Then, let us compute the filters in a recursive fashion. We first define

$$\begin{aligned}
H_0^1(z) &\equiv \beta \left(1 - \gamma_0 z^{-1}\right) \\
H_1^1(z) &\equiv \beta \left(\gamma_0 + z^{-1}\right)
\end{aligned}$$

and then use the recursion

$$\begin{aligned}
H_0^{i+1}(z) &= H_0^i(z) - \gamma_i z^{-2} H_1^i(z) \\
H_1^{i+1}(z) &= \gamma_i H_0^i(z) + z^{-2} H_1^i(z)
\end{aligned} \tag{3.35}$$

for all lattice sections, i.e., for $i = 1, \dots, N-1$. Then, the final filters will be

$$\begin{aligned}
H_0(z) &= H_0^N(z) \\
H_1(z) &= H_1^N(z)
\end{aligned}$$

Note that $H_1^1(z) = z^{-1} H_0^1(-z^{-1})$. It is also clear from (3.35) that if $H_0^i(z) = z^{-(2i-1)} H_0^i(-z^{-1})$, then

$$H_0^{i+1}(z) = z^{-(2i+1)} H_0^{i+1}(-z^{-1})$$

Thus, if at the ith stage we have CQF filters of length $2i$, then at the $(i+1)$th stage we have CQF filters of length $2i+1$. Because $H_0^1(z)$ and $H_1^1(z)$ are obviously CQF filters for $L = 2$, we conclude that the recursion in (3.35) will always generate CQF filters. Therefore, $H_0(z)$ and $H_1(z)$ will be CQF filters of length $2N$. In order that (3.31) and (3.32) be satisfied, the normalizing constant β should have the value [23]

$$\beta = \prod_{i=1}^{N-1} \sqrt{1 + \gamma_i^2}$$

It is interesting to note that not only the lattice structure will generate CQF filters, but the reciprocal is also true, that is, any CQF filter can be implemented by the lattice structure [23]. In order to see that, let us look at the inverse recursion corresponding to (3.35), which is

$$\begin{aligned}
H_0^i(z) &= (1 + \gamma_i^2)^{-1} \left[H_0^{i+1}(z) + \gamma_i H_1^{i+1}(z)\right] \\
H_1^i(z) &= (1 + \gamma_i^2)^{-1} z^2 \left[H_1^{i+1}(z) - \gamma_i H_0^{i+1}(z)\right]
\end{aligned} \tag{3.36}$$

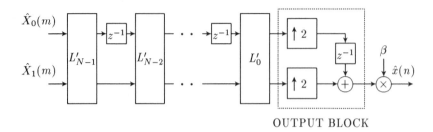

OUTPUT BLOCK

Figure 3.25. Lattice structure for perfect-reconstruction, two-channel filter banks, synthesis section.

From the above equation, we can actually obtain the parameter γ_i. Note that the order of $H_0^i(z)$ must be two less than the order of $H_0^{i+1}(z)$. Therefore, γ_i must be such that the coefficients in z^{-2i} and $z^{-(2i+1)}$ are cancelled in the right-hand side of the first equation in (3.36). Then, it follows that

$$\gamma_i = -\frac{h_0^{i+1}(2i+1)}{h_1^{i+1}(2i+1)} = -\frac{h_0^{i+1}(2i)}{h_1^{i+1}(2i)} \tag{3.37}$$

Note that the two forms above are only equivalent if $H_0^{i+1}(z)$ and $H_1^{i+1}(z)$ are a valid CQF pair. Thus, given a CQF pair $H_0(z)$ and $H_1(z)$ of length $L = 2N$, we can use (3.37) and (3.36), for $i = N-1, N-2, \ldots, 0$ to obtain the parameters γ_i of the lattice sections.

The synthesis lattice is shown in Fig. 3.25. It is quite similar to the analysis structure, but with the following difference: if γ_i is the parameter for the analysis lattice section L_i, the synthesis section L_i' will have parameter $-\gamma_i$. It is clear that the cascade connection of L_i and L_i' is an identity operator, within a scale factor. Therefore, when we cascade the analysis and synthesis filter banks, all lattice sections cancel out, and only the delays remain. This observation allows us to conclude immediately that the lattice structures will have the perfect reconstruction property. For a more detailed proof, see [23]. The above ideas are summarized in the CQF design procedure described below.

Method for the Design of CQF Filter Banks

Step 1: Pick up the filter length $L = 2N$.

Step 2: Design a half-band filter $u(n)$ with length $2L - 1$, using any suitable FIR filter design procedure that generates half-band filters (for example, the window method). Recall that $U(e^{j\omega})$ should be a low-pass filter with a cutoff frequency of $\pi/2$, and it should have approximately twice the stopband attenuation (in dB) that we want $H_0(e^{j\omega})$ to have. We may have to modify our initial guess of the filter length during this step in order to obtain the desired stopband attenuation.

Step 3: Set

$$u(0) \leftarrow u(0) - \delta_s$$

where $\delta_s = \min\{U(e^{j\omega})\}$.

Step 4: Compute the zeros of $U(z)$. They should occur in reciprocal pairs. Pick half of the zeros and call them $\{z_1, z_2, \ldots, z_{L-1}\}$. Note that this choice means that if z_k from the set is not on the unit circle then $1/z_k$ should *not* be in the set. Then, define $H_0(z)$ by

$$H_0(z) = \sqrt{G} \prod_{l=1}^{L-1} (z^{-1} - z_l)$$

The constant G should be chosen such that $\sum_{n=0}^{L-1} |h_0(n)|^2 = 1$. Define $H_1(z)$ by

$$H_1(z) = z^{-(L-1)} H_0(-z^{-1})$$

The filters $H_0(z)$ and $H_1(z)$ are a CQF pair of length L. Recall that, in order to avoid sharp discontinuities in the impulse responses (and consequently the blocking effects), we should pick up some of the zeros inside the unit circle, and some of them outside.

Step 5: Use equations (3.37) and (3.36), for $i = N - 1, N - 2, \ldots, 0$ to obtain the parameters γ_i of the lattice sections.

An advantage of this design method is that it can easily be carried out interactively within the MATLAB software package [25], for example, which has all the necessary built-in functions for easy computation of the steps above. So, no programming is necessary. We have included in Appendix A a table of CQF filter coefficients which may be adequate for most applications.

It is important to point out that, in Step 2 of the CQF design procedure, we could use the Parks-McClellan algorithm [7] to design an optimum equiripple filter $U(z)$. Then, $H_0(z)$ and $H_1(z)$ would also have equiripple responses.

Another technique for the design of CQF filters is to work with the lattice parameters γ_i directly. After an initial guess of their values, an optimization procedure can search for the parameters that will minimize an useful cost function, such as the stopband energy

$$E = \int_{\omega_s}^{\pi} |H_0(e^{j\omega})|^2 d\omega$$

This technique was suggested in [23]. The problem with this optimization procedure is that the objective function is highly nonlinear on the free parameter space, and convergence to a good solution is dependent on the initial choice of the parameters. More details of this approach can be found in [23].

It is possible to modify slightly the lattice structure of Fig. 3.23 so that the filters $H_0(z)$ and $H_1(z)$ will have linear phase [26]. Forcing linear phase means giving up the power-complementary property, as we have seen earlier. The main problem is that linear phase and perfect reconstruction implies in reduced stopband attenuations, as shown in [26]. For example, a linear-phase lattice with filters of length $L = 64$ can be designed so that the filters will have frequency responses similar to those of QMF filters of length 32. Because the lattice is more efficient to implement, both solutions end up with roughly the same computational complexity. The only disadvantage of the lattice structure is that it will have twice the input-to-output delay.

Note that either with the CQF lattice or with the linear-phase lattice we have perfect reconstruction. Thus, the cascade of analysis and synthesis filters for each subband is necessarily linear-phase. Thus, we do not believe that it is worthwhile to pay the price of reduced stopband attenuations with the linear-phase lattice. If we use a CQF lattice we can reduce the computational complexity by approximately 50 percent, for a given set of filter specifications (transition bandwidth and stopband attenuation). However, we must be careful to pick up spectral factors for $H_0(z)$ such as to avoid blocking effects. This means we should guarantee that $h_0(n)$ decays to zero smoothly at its extremities, as discussed before. The price we have to pay is a probably longer design time, because we may have to pick up a good spectral factor in a trial-and-error basis.

3.4.2 PR Filter Banks for $M > 2$

We have seen in the previous subsection that CQF filter banks are an extension of the QMF concept to achieve perfect reconstruction when $M = 2$. A logical question at this point is: could we extend the pseudo-QMF filter banks of Section 3.3 to obtain a PR filter bank for arbitrary M? The answer to this question is indeed affirmative, and in fact it will lead to the modulated lapped transforms, which we discuss in more detail in Chapter 5.

There was another approach toward the solution of the M-band PR FIR filter bank design problem [24]. Instead of starting from pseudo-QMF filter banks, the idea was to try to generalize the CQF concept to the M-band filter bank structure of Fig. 3.7. It is easier to analyze the problem in the frequency domain, and so let us rewrite (3.9) and (3.10) in the form [24]

$$\hat{X}(z) = \frac{1}{M} \sum_{l=0}^{M-1} X(zW_M^{-l}) \sum_{k=0}^{M-1} H_k(zW_M^{-l}) F_k(z) \tag{3.38}$$

where, as usual, $W_M \equiv e^{-j2\pi/M}$, with $j \equiv \sqrt{-1}$. The above equation can be put in matrix form as

$$\hat{X}(z) = \frac{1}{M} \mathbf{F}^T(z) \mathbf{H}^T(z) \mathbf{X}(z) \tag{3.39}$$

where $\mathbf{F}^T(z) \equiv [F_0(z) \ F_1(z) \ \cdots \ F_{M-1}(z)]$ is the vector composed of the responses of the synthesis filters, $\mathbf{X}^T(z) \equiv [X(z) \ X(zW_M^{-1}) \ \cdots \ X(zW_M^{-M+1})]$ is the vector with the input signal z-transform and its aliased versions, and

$$\mathbf{H}(z) \equiv \begin{bmatrix} H_0(z) & H_1(z) & \cdots & H_{M-1}(z) \\ H_0(zW_M^{-1}) & H_1(zW_M^{-1}) & \cdots & H_{M-1}(zW_M^{-1}) \\ \vdots & \vdots & & \vdots \\ H_0(zW_M^{-M+1}) & H_1(zW_M^{-M+1}) & \cdots & H_{M-1}(zW_M^{-M+1}) \end{bmatrix} \tag{3.40}$$

is the *aliasing component matrix*, or AC-matrix [9].

With the AC-matrix approach, it is quite clear what are the conditions for perfect reconstruction. All the aliasing components of $\mathbf{X}(z)$ should be removed, and the gain for $X(z)$ should be unity. Therefore, perfect reconstruction would require [9, 24]

$$\mathbf{H}(z)\,\mathbf{F}(z) = [Mz^{-D}\ 0\ 0\ \cdots\ 0]^T \tag{3.41}$$

where D is some acceptable delay. Thus, given a set of analysis filters $\{H_0(z)$ to $H_{M-1}(z)\}$, the set of equations in (3.41) could be inverted in order to obtain the synthesis filters as

$$\mathbf{F}(z) = \mathbf{H}(z)^{-1}\,[Mz^{-D}\ 0\ 0\ \cdots\ 0]^T \tag{3.42}$$

This approach, however, might lead to IIR synthesis filters, without any guarantee of stability, because $\mathbf{H}(z)^{-1}$ could introduce poles outside the unit circle [9]. Thus, direct inversion of the AC-matrix is not advisable. A set of time-domain conditions that are equivalent to (3.41) was derived in [27].

For practical reasons, we would like to use FIR filters both in the analysis and in the synthesis filter banks. The AC-matrix formulation is still useful, though, because (3.41) must be valid. For example, consider the PR solution that uses no filtering in the FB-II structure of Fig. 3.8. It corresponds, in the original FB-I structure, to the choice of filters

$$H_k(z) = z^{-k}, \quad F_k(z) = z^{M-1-k}$$

Although this is a trivial and useless solution, it satisfies the PR condition in (3.41), with $D = M - 1$. This trivial solution, shown in Fig. 3.26, is actually quite important, as first observed in [24], because it demonstrates the possibility of perfect reconstruction with FIR filters.

Note that, as we have already commented in Section 3.2 during our discussion of the DFT filter bank, the set of delays and M:1 decimators on the left of Fig. 3.26 are a multirate representation of the simple operation of gathering a block of M consecutive input samples. Similarly, the 1:M interpolators and delay elements on the right of Fig. 3.26 are a multirate representation of appending a block of M samples to the output. Thus, the structure of Fig. 3.26 just picks up a block of input samples and sends them to the output.

A major step in the generation of PR solutions with FIR filters was the recognition that if an invertible transfer function matrix is inserted in the structure of Fig. 3.26, then the PR property will be preserved. The basic idea, suggested by

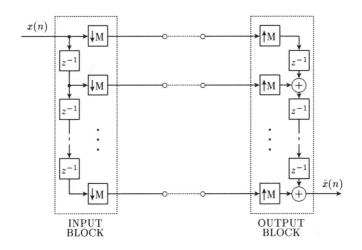

Figure 3.26. Trivial perfect reconstruction filter bank.

Vaidyanathan [24] as an extension of the polyphase structure for $M = 2$, is shown in Fig. 3.27.

The M-input M-output transfer function matrix $\mathbf{E}(z)$ is inserted in the analysis filter bank, and $z^{-(N-1)}\tilde{\mathbf{E}}(z)$ is inserted in the synthesis bank. Following [24], we define

$$\tilde{\mathbf{E}}(z) \equiv \mathbf{E}^T(z^{-1})$$

Because the decimators appear before $\mathbf{E}(z)$ in the analysis filter bank of Fig. 3.27, and comparing it with Fig. 3.7 we see that $\mathbf{E}(z)$ actually contains the polyphase components of the analysis filters, in the form [24]

$$H_k(z) = \sum_{r=0}^{M-1} z^{-r} E_{kr}(z^M) \tag{3.43}$$

where the subscript kr means the kth row and rth column. As usual throughout this book, we assume that the rows and columns are numbered from 0 to $M - 1$. It

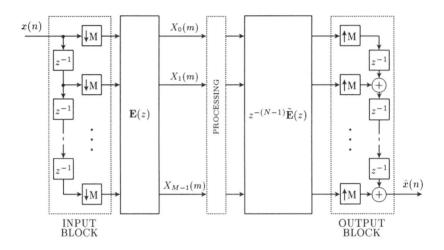

Figure 3.27. Perfect reconstruction filter bank based on the introduction of a lossless matrix $\mathbf{E}(z)$.

is also clear from Fig. 3.27 that the polyphase components of the synthesis filters are given by $\tilde{\mathbf{E}}(z)$. Thus, their polyphase decompositions are [24]

$$F_k(z) = \sum_{r=0}^{M-1} z^{-(NM-1-r)} \, \tilde{E}_{rk}(z^M) \tag{3.44}$$

Because of the relationships above, we will refer to $\mathbf{E}(z)$ as the *polyphase component matrix*.

Now, let us assume that $\mathbf{E}(z)$ is such that

$$\tilde{\mathbf{E}}(z) \, \mathbf{E}(z) = \mathbf{E}(z) \, \tilde{\mathbf{E}}(z) = \mathbf{I} \tag{3.45}$$

where $\mathbf{I}$ is the identity matrix. Then the polyphase component matrix $\mathbf{E}(z)$ is called *paraunitary*, or *lossless* [24]. The reason for the first term is that $\mathbf{E}(z)$ is unitary (orthogonal) in the unit circle $z = e^{j\omega}$. Thus, all eigenvalues of $\mathbf{E}(e^{j\omega})$ have unity magnitude, and therefore its "gain" is equal to one at all frequencies. This is an extension of the all-pass concept, and it is one of the reasons for the denomination lossless. Another reason is that such an $\mathbf{E}(z)$ has similar properties as passive

RLC structures [28]. Furthermore, lossless transfer functions have low passband sensitivities to variations in the multiplier coefficients. This is certainly a desirable feature in practice, where we must work with quantized coefficients.

When $\mathbf{E}(z)$ is lossless, it is clear that the cascade of $\mathbf{E}(z)$ and $z^{-(N-1)}\tilde{\mathbf{E}}(z)$ in Fig. 3.27 results just in a net delay of $N - 1$ samples for each decimated subband signal. Therefore, (3.41) and (3.42) hold, with a delay $D = NM - 1$. Generally we are interested in causal filters, and we also want a minimum delay for a given filter order. Therefore, it is clear that $z^{-(N-1)}\tilde{\mathbf{E}}(z)$ will have no positive powers of z when the order of $\tilde{\mathbf{E}}(z)$ is $N - 1$.

With a polyphase component matrix of order $N - 1$, it follows from (3.43) and (3.44) that $H_k(z)$ and $F_k(z)$ are FIR filters of length $L = NM$. Furthermore, (3.44) is equivalent to

$$F_k(z) = \sum_{r=0}^{M-1} z^{-(NM-1-r)} E_{kr}(z^{-M})$$

which implies

$$F_k(z) = z^{-(L-1)} H_k(z^{-1}) \tag{3.46}$$

or, in the time domain,

$$f_k(n) = h_k(L - 1 - n) \tag{3.47}$$

Then the impulse responses of the synthesis filters are equal to the time-reversed impulse responses of the corresponding analysis filters. This is an extension to arbitrary M of CQF filters. Note that (3.46) also implies that $|F_k(e^{j\omega})| = |H_k(e^{j\omega})|$, i.e., the analysis and synthesis filters for each subband have the same magnitude response.

Using (3.43), the AC-matrix $\mathbf{H}(z)$ of (3.40) assumes the form [24]

$$\mathbf{H}(z) = \sqrt{M}\,\mathbf{A}\mathbf{D}(z)\,\mathbf{E}^T(z^M)$$

where $\mathbf{A}$ is the orthogonal DFT matrix of order M defined in (1.15) and $\mathbf{D}(z)$ is a diagonal matrix of delays defined by $\mathbf{D}(z) \equiv \mathrm{diag}\,\{1, z^{-1}, \ldots, z^{-(M-1)}\}$. From the above equation and the losslessness of $\mathbf{E}(z)$, we conclude that

$$\tilde{\mathbf{H}}(z)\,\mathbf{H}(z) = M\,\mathbf{I} \tag{3.48}$$

and so $\sqrt{1/M}\,\mathbf{H}(z)$ is also lossless. Note that the above equation is equivalent to $\mathbf{H}(z)\,\tilde{\mathbf{H}}(z) = M\,\mathbf{I}$, which implies

$$\sum_{k=0}^{M-1} H_k(z^{-1})\,H_k(z) = M$$

or, in the Fourier domain $z = e^{j\omega}$

$$\sum_{k=0}^{M-1} |H_k(e^{j\omega})|^2 = M \qquad (3.49)$$

Thus, the analysis filters are power-complementary, an extension of the CQF property for $M = 2$ in (3.34). In view of (3.46), it is clear that the synthesis filters are also power-complementary.

Looking at the main diagonal of (3.48), we see that

$$\sum_{r=0}^{M-1} H_k(z^{-1}W_M^r)\,H_k(zW_M^{-r}) = M$$

If we define

$$U_k(z) \equiv H_k(z)\,H_k(z^{-1}) \qquad (3.50)$$

we conclude that

$$\sum_{r=0}^{M-1} U_k(zW_M^{-r}) = M \qquad (3.51)$$

Therefore, $U_k(z)$ is an Mth band filter [1], i.e.,

$$u_k(0) = 1, \quad \text{and} \quad u_k(rM) = 0,\ r = \pm 1, \pm 2, \pm(N-1)$$

Thus, each analysis filter is a spectral factor of an Mth band filter. This is also a generalization of the CQF filters, where for $M = 2$ the filters were spectral factors of a half-band filter. In the CQF case, though, $H_0(z)$ and $H_1(z)$ could be chosen as the spectral factors of the *same* filter. Unfortunately, this property does not extend to the case of arbitrary M, as shown in [24]. Therefore, the design technique for

CQF filters based on the spectral factorizations of a single filter cannot be used for arbitrary M.

From the above discussion, we see that we need new design techniques for FIR PR filter banks based on lossless transfer function matrices. In [24], it was suggested that the polyphase component matrix be composed of a cascade of zero-order lossless matrices (which are just orthogonal matrices) and diagonal matrices containing only delays. An orthogonal matrix can always be factored as a cascade of plane rotations, where each plane rotation is actually a sparse matrix containing only a 2×2 butterfly [24] characterized by a rotation angle. Thus, a possible design technique could be the use of a nonlinear optimization routine that would try to find an optimum set of rotation angles, with a cost function that can be, for example, the total stopband energy of all analysis filters. In view of the power-complementary property, if we get good stopband characteristics for all analysis filters, then their passband characteristics are automatically guaranteed to be good.

As we saw in Chapter 2, the number of degrees of freedom for an orthogonal matrix of order M is $M(M-1)/2$. Each plane rotation has one parameter, the rotation angle, and so the total number of angles to completely characterize any orthogonal matrix would be $M(M-1)/2$. The factorization of $\mathbf{E}(z)$ proposed in [24] may have up to N orthogonal matrices, for filters of length $L = NM$. Thus, the total number of rotation angles, in the most general structure, would be $NM(M-1)/2$. This represents a major problem if we want to design a filter bank with $M > 16$. For example, a 32-channel bank with filters of length 128 would have 1,984 free parameters. Because the cost function is highly nonlinear on the parameters, the optimization procedure might fail to converge even to a reasonably acceptable solution.

There has been a lot of research carried out by Vaidyanathan an colleagues on trying to make the optimization problem more robust. The main objectives were two: first, the reduction of the number of free parameters by means of other factorizations of $\mathbf{E}(z)$; second, the development of a systematic procedure to define a good set of initial values for those parameters. In [29], the parametrization was made more efficient by imposing the restriction of pairwise mirror-image symmetry in the frequency responses, that is, $H_{M-1-k}(z) = z^{-r}H_k(z^{-1})$ (for M even), which reduces significantly the number of free parameters. This restriction actually makes a lot of sense, in practice, because it forces the band-pass filters for the high-frequency subbands to be as good as the filters for the low-frequency subbands. A different parametrization of $\mathbf{E}(z)$, based on a state-space description, was proposed in [30].

This new structure is not only canonical (i.e., it can realize ant lossless transfer function matrix), but also it has a lower number of free parameters.

Returning to the factorization based on the cascade of orthogonal matrices and delays, it was later shown in [31] that the eigenfilter approach [32] could be used to design the Mth band filter $U_0(z)$ in (3.50) in such a way as to guarantee the existence of a spectral factor $H_0(z)$ with controllable filtering characteristics. Once $H_0(z)$ is fixed, the number of free rotation angles is significantly reduced, and in [31] it was shown that there is a synthesis procedure that can initialize most of the free parameters to good initial values. With this new approach, not only the time consumed in the optimization was drastically reduced, but the final results were much better than those obtained with the previous design techniques. For example, filters for $M = 3$ with stopband attenuations in excess of 70 dB were reported in [31].

Further design simplification can be obtained by forcing the analysis filters to have linear phase [33]. For that, we must give up the losslessness of the polyphase matrix $\mathbf{E}(z)$ and, therefore, the power-complementary property, but still keeping perfect reconstruction. This has the disadvantage of leading to synthesis filters with impulse responses that might be longer than those of the analysis filters. The advantage is that the filter banks can be implemented by efficient lattice structures, in which the PR property is not lost by quantization of the multiplier coefficients.

We conclude this section with the observation that even with the improved techniques discussed above, the design of perfect-reconstruction FIR filter banks is far from being an easy task, mainly for large M, e.g., $M > 16$. Furthermore, the computational complexity of these general filter banks grow proportional to M^2, because of the general orthogonal matrices that are factors of $\mathbf{E}(z)$. This is an important point, because there are practical application of subband signal processing in which we would like to use a large number of subbands, as discussed in Section 3.6.

When we recall from Chapter 2 that with block transforms we had a computational complexity proportional to $\log_2 M$, a natural question that arises is: are there restricted structures for the PR FIR filter banks such that their computation can be based on fast transforms? If the answer is affirmative, then we could have PR FIR filter banks with complexities similar to those of block transforms, even for large M. Fortunately, the answer is affirmative, and it is based on the fact that the pseudo-QMF filter bank can be designed such as to force the PR property. This

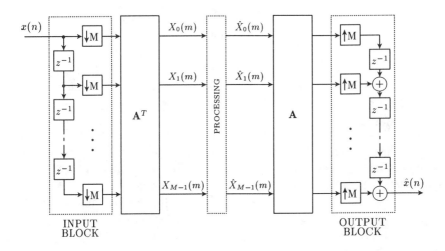

Figure 3.28. Perfect reconstruction filter bank with a zero-order lossless matrix $\mathbf{E}(z) = \mathbf{A}^T$.

is the basis for the modulated lapped transform (MLT), which we will discuss in Chapter 5.

3.5 Block Transforms versus Filter Banks

It should be clear that there are many more similarities than differences between signal processing with block transforms and with perfect-reconstruction multirate filter banks. This fact is well known [1], and the similarity can very easily be seen with the help of the structures based on lossless polyphase component matrix. With the basic PR filter bank structure of Fig. 3.27, suppose we choose a zero-order polyphase component matrix, in the form

$$\mathbf{E}(z) = \mathbf{A}^T \tag{3.52}$$

where $\mathbf{A}$ is an orthogonal matrix. Any orthogonal matrix is also lossless [see (3.45)], and thus the choice of $\mathbf{E}(z)$ in (3.52) leads to a PR filter bank, which is shown in Fig. 3.28. Note that $\tilde{\mathbf{E}}(z) = \mathbf{A}$.

Now, consider what is actually being done to the input signal in Fig. 3.28. On the left, the delay units and $M : 1$ decimators serve to gather a M-sample input block. This block is transformed by the zero-memory operator $\mathbf{A}^T$ to generate the subband signals. After processing these subband signal, the inverse transform $\mathbf{A}$ is applied, and on the right of Fig. 3.28 the resulting block of M samples is appended to the output. This is precisely what we did in Chapter 2 for signal processing with block transforms.

Therefore, we conclude that signal processing with a block transform, as discussed in Chapter 2, is a special case of signal processing in subbands with a perfect-reconstruction FIR filter bank. Furthermore, we could put the $M : 1$ decimators in Fig. 3.28 after the transform $\mathbf{A}^T$, and the $1 : M$ interpolators before the inverse transform $\mathbf{A}$ (recall from Section 3.1 that memoryless operators commute with decimators or interpolators). Then, we would immediately see that the basis functions of the transform (the columns of $\mathbf{A}$) would be the impulse responses of the synthesis filters, and the time-reversed basis functions would be the impulse responses of the analysis filters [see (3.47)].

With the lossless polyphase component matrix formulation, we can immediately see some fundamental properties of block transforms. First, their basis functions are filters that satisfy the power-complementary property in (3.49). Furthermore, we see that these filters cannot have a good stopband performance because their lengths is only M for a bandwidth of π/M. For example, the frequency responses of the DCT basis functions [34] are shown in Fig. 3.29, from which we see that not only their bandwidths are significantly larger than π/M, but the stopband attenuation is only about 10 dB in the worst case.

The subband signals with the DCT filter bank will contain high levels of aliasing, due to the poor filtering performed by the DCT basis functions. This aliasing is cancelled in the synthesis banks, according to the PR property, but only if the subband signals are *not* modified, which is seldom the case in practice. In adaptive filtering or signal coding, each subband signal will be subject to noise and, at least, a different gain in the processing. Therefore, the aliasing generated in the analysis filters will not be cancelled anymore. This uncanceled aliasing will show up in the time domain as the *blocking effects* that we discussed in Chapter 2. This is particularly true in low bit rate transform coding, where the higher frequency subbands are not transmitted at all, and so the aliasing cancellation in the synthesis filter banks cannot take place.

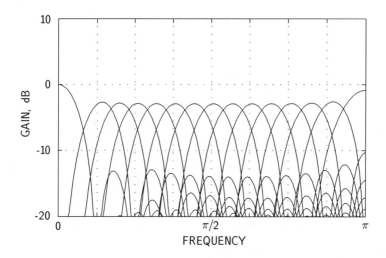

Figure 3.29. Magnitude responses of the DCT filter bank for $M = 16$.

Thus, it is clear that in order to reduce the blocking effects, we must have analysis filters with a low level of aliasing, that is, good filtering characteristics. This is why a PR filter bank designed with long FIR filters does not lead to blocking effects, even at very low bit rates (as long as the impulse responses of those filters decay smoothly to zero at their ends, as discussed in the design of CQF filters). Hence, as long as we can tolerate the further delay in processing, it would be better to use a PR filter banks than a block transform. The reason why this is not always done in practice is the computational complexity of FIR filter banks. Lapped transforms, however, are perfect-reconstruction filter banks with much better filtering characteristics than block transforms, and a computational complexity close to that of block transforms. Therefore, with lapped transforms we can efficiently implement subband signal processing systems for a large number of subbands.

From the above discussion, we concluded that block transforms can be viewed as M-channel filter banks with filters of length M. Within the same viewpoint, a general M-channel perfect-reconstruction filter bank can be considered as an *extended transform*, in which the basis functions as longer than the block size, and the original orthogonality condition of the basis functions is replaced by the more general losslessness condition of $\mathbf{E}(z)$. This is precisely the reasoning that went behind the development of lapped transforms, as we will discuss in more detail in Chapters 4 and 5.

3.6 Applications

Given the similarities between block transforms and multirate filter banks, we should expect that the applications of subband signal processing should be the same as those of block transforms. This is indeed so, and in all applications that we discussed in Chapter 2 multirate filter banks can also be successfully employed. Some examples are:

- Signal filtering

- Spectrum estimation

- Signal coding

- Transmultiplexing

As we discussed in Chapter 2, the main advantage of using block transforms in those applications was the fast algorithms associated with most of the sinusoidal transforms. Thus, if we want to replace the transforms by multirate filter banks, it would be desirable that the filter banks could also be computed by fast algorithms. In this sense, in most applications lapped transforms can be very attractive because they have algorithms with complexities close to those of block transforms, and they have good subband filtering characteristics. Thus, in this section we will just review briefly some of the applications, and in Chapter 7 we will discuss in detail the applications of multirate filter banks based on lapped transforms.

3.6.1 Signal Filtering

In Chapter 2 we discussed the overlap-add and overlap-save techniques for efficient FIR filtering based on block transforms. These techniques, although realizing FIR filtering exactly, have the disadvantage of having to work with zero-padded signal blocks. If we use multirate filter banks, no zero padding is necessary. The basic idea is shown in Fig. 3.30, which is a particular case of the FB-I structure of Fig. 3.7, in which the processing of the subband signals is just the multiplication of each

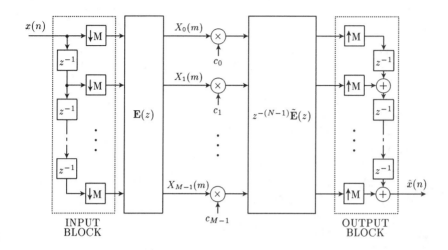

Figure 3.30. FIR filtering with multirate filter banks.

subband by a different gain. In [35] it was shown that by appropriate choices of the analysis and synthesis filters, the structure of Fig. 3.30 could be equivalent to a single shift-invariant FIR or IIR filter. For this, the multipliers in Fig. 3.30 should be replaced by appropriate transfer functions, which would have lower order than the original FIR or IIR filter to be implemented.

The structure in Fig. 3.30 is quite similar to the FDAF-II structure of Fig. 2.6. The only difference is that in Fig. 3.30 the weights are not automatically adjusted according to some desired response. By changing the weights in Fig. 3.30, we can actually have a variable filter, which in audio systems is usually referred to as a *graphic equalizer*. If a perfect reconstruction filter bank based is used in Fig. 3.30, we know that we would recover the original signal $x(n)$ exactly, within a delay, if we set all c_k to one. If all c_k are not equal, though, the mechanism of aliasing cancellation would not work, and the structure of Fig. 3.30 would not be equivalent to shift-invariant FIR filtering. Nevertheless, if the analysis filters have good stopband attenuations and narrow transition widths, then the amount of aliasing introduced in the analysis filter bank would be low, and the residual error due to uncanceled aliasing might be tolerable, depending on the application.

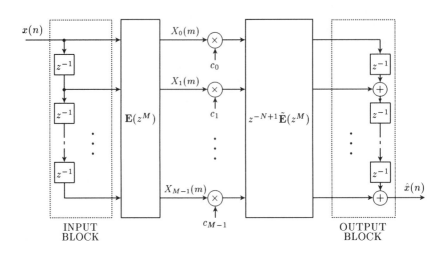

Figure 3.31. Graphic equalizer with single-rate filter banks.

In high-fidelity audio, for example, where even a 70 dB signal-to-noise ratio may be considered low, we could use the structure of Fig. 3.31 to implement a graphic equalizer. In that structure, we have removed the decimators and interpolators, and so we have a single-rate system that is linear and shift-invariant. Thus, there is no aliasing to be cancelled. Note that we have used $\mathbf{E}(z^M)$ in Fig. 3.31 because there are no decimators.

Let us assume that we use perfect reconstruction filter banks based on lossless transfer function matrices in Fig. 3.31. Then the overall transfer function $G(z) \equiv \hat{X}(z)/X(z)$ is given by

$$
\begin{aligned}
G(z) &= \sum_{k=0}^{M-1} c_k\, H_k(z)\, F_k(z) \\
&= z^{-(L-1)} \sum_{k=0}^{M-1} c_k H_k(z)\, H_k(z^{-1})
\end{aligned}
$$

or, in the Fourier domain

$$
G(e^{j\omega}) = e^{-j\omega(L-1)} \sum_{k=0}^{M-1} c_k\, |H_k(e^{j\omega})|^2 \tag{3.53}
$$

Due to the power-complementary property of PR filter banks based on lossless matrices, we have, if all the gains are equal

$$c_k = G, \; \forall k \; \implies \; G(e^{j\omega}) = G, \; \forall \omega$$

Thus, the graphic equalizer would just delay the signal, with a constant gain and a linear phase response.

If the filter banks in Fig. 3.31 could be implemented with fast transforms, that structure would be computationally more efficient than an equalizer based on independent FIR filters. This is the case if the filter banks are lapped transforms, as we see in Chapter 7.

3.6.2 Adaptive Filtering

A filter bank can replace a block transform in the FDAF-I or FDAF-II structures of Section 2.1. We recall that the purpose of the transform in the adaptive filter structures was to convert a strongly colored signal into a set of approximately white subband signals. adaptive In this way, the LMS algorithm would have a much faster convergence when dealing with the subband signals, as demonstrated in Fig. 2.7. We have seen in Section 3.5 that filter banks can do a much better job of separating the input spectrum into its subbands than block transforms. Thus, we would expect that the performance of the frequency domain adaptive filters could be improved by using filter banks instead of block transforms. This is indeed true, as shown in [36].

If a multirate filter bank is used in the FDAF-I structure, the variations in the adjustable weights would lead to a lack of aliasing cancellation. As pointed out in [36], this aliasing will occur mainly between each pair of adjacent subbands. Thus, it is possible to introduce adaptive cross-coupling weights in Fig. 2.5 such that adjacent channel aliasing continues to be cancelled, for any value of the adaptive weights. In many applications, for example modems, a small amount of uncanceled aliasing may not be a problem, so that the original FDAF-I structure can be used.

3.6.3   Spectrum Estimation

The basic technique that we discussed in Chapter 2 for spectrum estimation using transforms was the averaging of periodograms. We could compute each periodogram from an analysis filter bank, and then build a spectrum estimator in the form

$$S_{xx}(e^{j\omega_k}) \simeq \frac{1}{N_B} \sum_{m=0}^{N_B-1} |X_k(m)|^2 \qquad (3.54)$$

where the frequency grid is $\omega_k = (k + 1/2)\pi/M$, for $k = 0, 1, \ldots, M - 1$. The estimator is built as the average of the periodograms of N_B consecutive blocks.

For applications where the spectrum estimator will be used to find narrowband spectral components in the incoming signal, the estimator in (3.54) will have a better resolution if a multirate filter bank is used instead of a block transform to compute $X_k(m)$. This is because of the better band-pass responses of the filters in the bank, compared to the responses of the transform basis functions. The price to be paid with multirate filter banks is the increase in computational complexity, but with lapped transforms this penalty can be relatively small, mainly for large M.

3.6.4   Signal Coding

In the classical transform coding system of Fig. 2.10, we can replace the transform operators by multirate filter banks. We would then be performing *subband coding*, which has been successfully applied to speech [37] and images [38, 39]. One of the main advantages of subband coding is the absence of the blocking effects that are characteristic of transform coding, as we discussed in Section 2.3. As in transform coding, subband coding leads to better results if the quantizers are adaptive [40].

Generally, subband coding has meant coding with a small number of subbands, because of the computational complexity of the filter bank. Thus, special adaptation rules have been derived for subband coders, with a bit assignment depending not only on the channel, but also varying with time [40], because of the nonwhiteness of the subband signals. With lapped transforms, we can actually perform subband coding with a large number of subbands, keeping the computational complexity low enough for easy real-time implementation. We will return to subband coding with lapped transforms in Chapter 7.

3.6.5 Other Applications

There are many other applications where multirate filter banks can be employed, for subband signal processing. For example, in speech scrambling [41], the block labeled "processing" in the FB-I structure of Fig. 3.7 could be simply a permutation matrix. The reconstructed signal $\hat{x}(n)$ would have its frequency components scrambled with respect to the original signal $x(n)$. To an eavesdropper, $\hat{x}(n)$ would be unintelligible, mainly if the permutation matrix varies with time in a pseudorandom manner. In a secure speech communication system, the transmitter would use an analysis-synthesis filter bank with some permutation matrix, and the receiver would use another analysis-synthesis filter bank, with the inverse permutation matrix. For this application, the use of block transforms would lead to blocking effects that would be even more disturbing than in transform coding because the analysis/synthesis operation is performed twice.

Finally, another classical application of multirate filter banks is in TDM-FDM transmultiplexing [42]. In this case we have a set of M signals $x_0(m)$ to $x_{M-1}(m)$, sampled at f_s, that we would like to combine, by frequency translations, into a single signal $y(n)$ sampled at Mf_s. The signal $y(n)$ would be transmitted, and from it we would like to recover the M original signals. The first step (frequency multiplexing) can be performed by an analysis filter bank, and the second step (frequency demultiplexing) can be performed by a synthesis filter bank. Because the filters are not ideal, there will be crosstalk among the channels in the multiplexed signal $y(n)$. However, if the analysis and synthesis filters are PR, then the crosstalk will be cancelled at demultiplexing. This is because, as shown in [42], subband processing and transmultiplexing are dual systems, so that aliasing in the first system corresponds to crosstalk in the second.

The main advantage of using perfect-reconstruction filter banks in transmultiplexing is that no guard bands are necessary. Thus, if we have 32 speech channels sampled at 8 kHz, we could frequency-multiplex them into a single signal at a sampling frequency of 256 kHz. The original signals could then be exactly recovered from it with a FIR filter bank. If the filter bank is a lapped transform, perfect transmultiplexing is computationally more efficient than with the more general PR filter banks that we have discussed in the previous section.

3.7 Summary

In this chapter we reviewed the theory and design techniques for multirate filter banks, which are the fundamental component of subband signal processing systems. We started in Section 3.1 with a review of the basic concept of multirate signal processing, with emphasis on the properties of decimators and interpolators. In Section 3.2 we looked at the general filter bank structure for subband signal processing, with the equivalent forms FB-I and FB-II. We briefly discussed the time-domain conditions for perfect reconstruction of the signal after the cascade of analysis, decimation, interpolation, and synthesis. We also looked at the DFT filter bank, which was the first step towards the realization of a filter bank with low computational complexity. We saw that perfect reconstruction with FIR analysis and synthesis filters in the DFT filter bank is not possible.

In Section 3.3 we discussed in detail the quadrature mirror filter (QMF) concept. With two-channel ($M = 2$) banks, QMF filters allow complete cancellation, in the synthesis filter, of the aliasing generated in the analysis filter. We also discussed in detail the pseudo-QMF filter bank, which is a generalization of the QMF concept for $M > 2$. Pseudo-QMF filter banks do not lead, in general, to complete aliasing cancellation. However, they are based on the reasonable assumption that virtually all of the aliasing is generated between adjacent channels, and such aliasing is completely cancelled. As it was the case with DFT filter banks, pseudo-QMF filter banks have efficient implementations, based on fast transforms. The pseudo-QMF family is quite important, because it was recently found that it can generate perfect reconstruction filter banks for arbitrary M and for arbitrary length filters. This led to the development of the modulated lapped transform, to be discussed in Chapter 5.

The perfect reconstruction (PR) problem was studied in Section 3.4. In most applications, mainly when the allowed signal distortion is very low (e.g., in music coding), it is important that the filter banks do not introduce any signal distortion. Thus, in the absence of further processing of the subband signals, the synthesis bank should be able to recover a delayed version of the input signal exactly. We focused only on banks of FIR filters, because they are inherently stable, easy to implement, and can be designed such that the overall analysis-synthesis filter bank has a linear phase response.

We started our discussion with the two-channel case, where conjugate quadrature filters (CQF) emerged as the PR equivalent to the QMF bank. We saw that CQF filters can be easily designed by using a spectral factorization technique. Then we moved into the general problem of M-band PR FIR filter banks. We saw that a trivial solution to this problem is the FB-II structure with no filtering. However, by introducing a lossless FIR polyphase component matrix $\mathbf{E}(z)$ after the decimators in the FB-II structure, and the matrix $z^{-(N-1)}\mathbf{E}^T(z^{-1})$ before the interpolators, we could generate FIR filter banks with perfect reconstruction. The fundamental characteristics of PR systems based on lossless matrices, namely the power-complementary property and the equality of the analysis and synthesis filter impulse responses within time reversal, were emphasized.

We briefly discussed the design techniques for PR FIR filter banks, all based in optimization techniques. In all of these techniques, however, the number of free parameters and the computational complexity for running the filter bank grow proportionally to the square of the number of bands. Therefore, we believe that PR FIR filter banks with special structures should be preferred in practice, mainly for large M. Lapped transforms come as a compromise between generality of the filtering characteristics and easy of design and implementation, as we will see in the next chapters.

In Section 3.5 we discussed the relationships between block transforms and filter banks. It became clear that signal processing with a block transform is actually a special case of subband signal processing with a multirate filter bank, in which the filter impulse responses are the transform basis functions. Within this viewpoint, we can see the blocking effects as an excessive level of uncanceled aliasing, caused by the poor filtering characteristics of the transform basis functions.

The applications of subband coding systems based on multirate filter banks were briefly discussed in Section 3.6. Many examples of some of these applications will be discussed in detail in Chapter 7, with the filter banks being composed of lapped transforms.

References

[1] Crochiere, R. E., and L. R. Rabiner, *Multirate Digital Signal Processing*, Englewood Cliffs, NJ: Prentice-Hall, 1983.

[2] Vaidyanathan, P. P., "Quadrature mirror filter banks, M-band extensions, and perfect reconstruction techniques," *IEEE ASSP Magazine*, vol. 4, July 1987, pp. 4-20.

[3] Malvar, H. S., *Optimal Pre- and Post-filters in Noisy Sampled-Data Systems*, PhD thesis, Mass. Inst. Tech., Cambridge, MA, Sep. 1986.

[4] Malvar, H. S., and D. H. Staelin, "Optimal FIR pre- and postfilters for decimation and interpolation of random signals," *IEEE Trans. Commun.*, vol. 36, Jan. 1988, pp. 67–74.

[5] Vaidyanathan, P. P., "Multirate digital filters, filter banks, polyphase networks, and applications: a tutorial," *Proc. IEEE*, vol. 78, Jan. 1990, pp. 56–93.

[6] Malvar, H. S., "Modulated QMF filter banks with perfect reconstruction," *Electron. Lett.*, vol. 26, no. 13, June 1990, pp. 906–907.

[7] IEEE DSP Committee, ed., *Programs for Digital Signal Processing*, New York: IEEE Press, 1979.

[8] Ramstad, T., "IIR filterbank for subband coding of images," *IEEE Intl. Symp. Circuits Syst.*, Espoo, Finland, pp. 827–830, June 1988.

[9] Smith, M. J. T., and T. P. Barnwell III, "A new filter bank theory for time-frequency representation," *IEEE Trans. Acoust., Speech, Signal Processing*, vol. ASSP-35, March 1987, pp. 314–327.

[10] Swaminathan, K., and P. P. Vaidyanathan, "Theory and design of uniform DFT, parallel, quadrature mirror filter banks," *IEEE Trans. Circuits Syst.*, vol. CAS-33, Dec. 1986, pp. 1170–1191.

[11] Smith, M. J. T., and T. P. Barnwell III, "Exact reconstruction techniques for tree-structured subband coders," *IEEE Trans. Acoust., Speech, Signal Processing*, vol. ASSP-34, June 1986, pp. 434–441.

[12] Croisier, A., D. Esteban, and C. Galand, "Perfect channel splitting by use of interpolation, decimation, and tree decomposition techniques," *Proc. Int. Conf. Inform. Sci. Syst.*, Patras, Greece, pp. 443–446, Aug. 1976.

[13] Johnston, J. D., "A filter family designed for use in quadrature mirror filter banks," *IEEE Intl. Conf. Acoust., Speech, Signal Processing*, Denver, CO, pp. 291–294, Apr. 1980.

[14] Jain, V. K., and R. E. Crochiere, "Quadrature mirror filter design in the time domain," *IEEE Trans. Acoust., Speech, Signal Processing*, vol. ASSP-32, Apr. 1984, pp. 353–361.

[15] Nussbaumer, H. J., "Pseudo-QMF filter bank," *IBM Tech. Disclosure Bull.*, vol. 24, Nov. 1981, pp. 3081–3087.

[16] Rothweiler, J. H., "Polyphase quadrature filters – a new subband coding technique," *IEEE Intl. Conf. Acoust., Speech, Signal Processing*, Boston, MA, pp. 1280–1283, March 1983.

[17] Nussbaumer, H., and M. Vetterli, "Computationally efficient QMF filter banks," *IEEE Intl. Conf. Acoust., Speech, Signal Processing*, San Diego, CA, pp. 11.3.1–11.3.4, March 1984.

[18] Masson, J., and Z. Picel, "Flexible design of computationally efficient nearly perfect QMF filter banks," *IEEE Intl. Conf. Acoust., Speech, Signal Processing*, Tampa, FL, pp. 541–544, March 1985.

[19] Chu, P., "Quadrature mirror filter design for an arbitrary number of equal bandwidth channels," *IEEE Trans. Acoust., Speech, Signal Processing*, vol. ASSP-33, Feb. 1985, pp. 203–218.

[20] Cox, R. V., "The design of uniformly and nonuniformly spaced pseudoquadrature mirror filters," *IEEE Trans. Acoust., Speech, Signal Processing*, vol. ASSP-34, Oct. 1986, pp. 1090–1096.

[21] Malvar, H. S., "Pseudo lapped orthogonal transform," *Electron. Lett.*, vol. 25, no. 5, March 1989, pp. 312–314.

[22] Mintzer, F., "Filters for distortion-free two-band multirate filter banks," *IEEE Trans. Acoust., Speech, Signal Processing*, vol. ASSP-33, June 1985, pp. 626–630.

[23] Vaidyanathan, P. P., and P. Q. Hoang, "Lattice structures for optimal design and robust implementation of two-channel perfect-reconstruction QMF banks," *IEEE Trans. Acoust., Speech, Signal Processing*, vol. 36, Jan. 1988, pp. 81–94.

[24] Vaidyanathan, P. P., "Theory and design of M-channel maximally decimated quadrature mirror filters with arbitrary M, having the perfect reconstruction property," *IEEE Trans. Acoust., Speech, Signal Processing*, vol. ASSP-35, Apr. 1987, pp. 476–492.

[25] Little, J., and C. Moler, *A preview of* PC-MATLAB, Portolla Valley, CA: The Mathworks, Inc., 1989.

[26] Nguyen, T. Q., and P. P. Vaidyanathan, "Two-channel perfect reconstruction FIR QMF structures which yield linear-phase analysis and synthesis filters," *IEEE Trans. Acoust., Speech, Signal Processing*, vol. 37, May 1989, pp. 676–690.

[27] Vetterli, M., and D. Le Gall, "Perfect reconstruction FIR filter banks: some properties and factorizations," *IEEE Trans. Acoust., Speech, Signal Processing,* vol. 37, July 1989, pp. 1057–1071.

[28] Vaidyanathan, P. P., and S. K. Mitra, "Low passband sensitivity digital filters: a generalized viewpoint and synthesis procedures," *Proc. IEEE,* vol. 72, Apr. 1984, pp. 404–423.

[29] Nguyen, T. Q., and P. P. Vaidyanathan, "Maximally-decimated perfect reconstruction FIR filter banks with pairwise mirror-image analysis (and synthesis) frequency responses," *IEEE Trans. Acoust., Speech, Signal Processing,* vol. 36, May 1988, pp. 693–706.

[30] Doğanata, Z., P. P. Vaidyanathan, and T. Q. Nguyen, "General synthesis procedures for FIR lossless transfer matrices, for perfect-reconstruction multirate filter bank applications," *IEEE Trans. Acoust., Speech, Signal Processing,* vol. 36, Oct. 1988, pp. 1561–1574.

[31] Vaidyanathan, P. P., T. Q. Nguyen, T. Saramäki, and Z. Doğanata, "Improved technique for design of perfect reconstruction FIR QMF banks with lossless polyphase matrices," *IEEE Trans. Acoust., Speech, Signal Processing,* vol. 37, July 1989, pp. 1042–1056.

[32] Vaidyanathan, P. P., and T. Q. Nguyen, "Eigenfilters: a new approach to least-squares FIR filter design and applications including nyquist filters," *IEEE Trans. Circuits Syst.,* vol. CAS-34, Jan. 1987, pp. 11–23.

[33] Nguyen, T. Q., and P. P. Vaidyanathan, "Structures for M-channel perfect-reconstruction FIR QMF banks which yield linear-phase analysis filters," *IEEE Trans. Acoust., Speech, Signal Processing,* vol. 38, March 1990, pp. 433–446.

[34] Malvar, H. S., "The LOT: a link between block transform coding and multirate filter banks," *IEEE Intl. Symp. Circuits Syst.,* Espoo, Finland, pp. 835–838, June 1988.

[35] Vetterli, M., "Running FIR and IIR filtering using multirate filter banks," *IEEE Trans. Acoust., Speech, Signal Processing,* vol. 36, May 1988, pp. 730–738.

[36] Gilloire, A., and M. Vetterli, "Adaptive filtering in subbands," *IEEE Intl. Conf. Acoust., Speech, Signal Processing,* New York, pp. 1572–1575, Apr. 1988.

[37] Derby, J. H., and C. Galand, "Multirate subband coding applied to digital speech interpolation," *IEEE Trans. Acoust., Speech, Signal Processing,* vol. ASSP-35, Dec. 1987, pp. 1684–1698.

[38] Woods, J. W., and S. D. O'Neil, "Subband coding of images," *IEEE Trans. Acoust., Speech, Signal Processing*, vol. ASSP-34, Oct. 1986, pp. 1278–1288.

[39] Smith, M. J. T., and S. L. Eddins, "Analysis/synthesis techniques for subband image coding," *IEEE Trans. Acoust., Speech, Signal Processing*, vol. 38, Aug. 1990, pp. 1446–1456.

[40] Honda, M., and F. Itakura, "Bit allocation in time and frequency domains for predictive coding of speech," *IEEE Trans. Acoust., Speech, Signal Processing*, vol. ASSP-32, June 1984, pp. 465–473.

[41] Cox, R. V., D. E. Bock, K. B. Bauer, J. D. Johnston, and J. H. Snyder, "The analog voice privacy system," *AT&T Tech. J.*, vol. 66, Jan.-Feb. 1987, pp. 119–131.

[42] Vetterli, M., "A theory of multirate filter banks," *IEEE Trans. Acoust., Speech, Signal Processing*, vol. ASSP-35, March 1987, pp. 356–372.

Chapter 4

Lapped Orthogonal Transforms

Our goal in this chapter is to study in detail the theory of lapped transforms (LTs), their properties, and the relationships between LTs and multirate filter banks. As we have already discussed in Section 1.3, the driving force behind the development of the lapped orthogonal transform (LOT) was the reduction of blocking effects in transform coding. In Section 4.1 we see that the time-domain restrictions that were imposed to LTs in order to reduce the blocking effects are actually equivalent to forcing the corresponding polyphase component matrix to be lossless.

We should note that all LTs to be considered in this book are orthogonal, that is, their basis functions are orthogonal and have their norms equal to one. Historically, however, the term LOT has been used for a lapped transform whose basis functions have even or odd symmetry, and lengths equal to $L = 2M$, where M is the transform size (or the number of subbands). Thus, an LT is a general lapped transform with orthogonal basis functions, whereas an LOT is a special case of an LT, with $L = 2M$. In Section 4.2 we review the design techniques and the main properties of LOTs.

The significant reduction in the blocking effects that can be achieved with LTs, when compared to standard block transforms, is certainly an attractive feature. However, the LOT would not be successful in practice if it were not for its fast computational algorithm. Thus, in Section 4.3 we will review the fast LOT implementation for a small number of subbands. A fast LOT algorithm that can be used for any M is presented in Section 4.4. Computer programs for the two versions of the fast LOT are given in Appendix C.

Finally, we compare the coding performance of the various LOTs to that of the DCT in Section 4.5. The main point is that LOTs have higher coding gains than the DCT, not only for simple AR(1) signals, but also for spectra derived from real-

---

world signals. This increase in coding gain comes as a bonus, giving that the main advantage of the LOT is the reduction of blocking effects.

4.1   Theory of Lapped Transforms

As we have seen in Chapter 2, in signal processing with standard block transforms we divide the incoming signal into blocks of M samples. These samples are transformed by an orthogonal matrix of order M, where M is the block size, or the number of transform coefficients.

With a lapped transform, we map an L-sample input block into M transform coefficients, where L is the length of the LT basis functions, with $L > M$. Since we want to keep the total sampling rate, we compute M new transform coefficients for every new M input samples. Thus, after the computation of the LT of a signal block, we shift into the L-sample input buffer the M samples of the new block. In this way, there will be an overlap of $L - M$ samples in the computation of consecutive LT blocks.

These ideas can be visualized with the help of Fig. 4.1, where we have assumed, for simplicity, that $L = 2M$. The LT basis functions are the columns of an $L \times M$ matrix $\mathbf{P}$. Hence, $\mathbf{P}^T$ is the direct transform operator, whereas $\mathbf{P}$ is the inverse transform operator. Because of the overlapping of the direct transform operators, signal reconstruction with a lapped transform requires overlapping of the outputs of each inverse LT, as shown in Fig. 4.1. In the overlapping regions, the reconstructed samples from different blocks are simply added.

A distinguishing feature of lapped transforms is that, when we apply an inverse LT that to a set of transform coefficients, we obtain L samples that are not equal to the original L samples that were used to compute the direct LT. Therefore, it does not make sense to talk about the inverse LT of a single block. Only after the overlap of consecutive inverse LT blocks do we recover the original signal.

In Fig. 4.1 we have also assumed causal processing. In the first direct LT operator, for example, $x(0)$ is the most recent input sample, and so in the inverse transform for that block the first output sample occurs at $\hat{x}(0)$. The overall delay is $L - 1$ samples.

Using the multirate operators that we have discussed in Chapter 3, we can build the general block diagram of signal processing with lapped transforms, which is

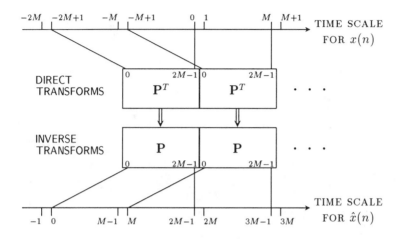

Figure 4.1. Signal processing with a lapped transform, for $L = 2M$. In the inverse transform, the reconstructed signal is formed by the superposition of the overlapping blocks, as in the overlap-add procedure discussed in Chapter 2.

shown in Fig. 4.2. Note that the input and output buffers in Fig. 4.2 are slightly different from those in the PR filter bank in Fig. 3.27. This is because, as it is usual in block processing, increasing indices of the transform basis functions correspond to an increase in the time index, whereas an increase in the polyphase component index corresponds to a decrease in the time index.

According to Figs. 4.1 and 4.2, the LOT of a signal block $\mathbf{x}$ can be obtained by

$$\mathbf{X} = \mathbf{P}^T \mathbf{x} \tag{4.1}$$

where $\mathbf{x}$ is an extended signal block having L samples

$$\mathbf{x} = [x(mM - L + 1) \quad x(mM - L + 2) \quad \ldots \quad x(mM - 1) \quad x(mM)]^T \tag{4.2}$$

with m being the block index. Thus, it is clear that each sample of the input signal $x(n)$ will be used in several blocks. Note that we have made implicit the dependence of $\mathbf{x}$ and $\mathbf{X}$ on m, as in Chapter 1, to simplify the notation.

With a traditional block transform $\mathbf{A}$, we saw in Chapter 2 that orthogonality of $\mathbf{A}$ guarantees perfect reconstruction in the inverse transform. With a lapped transform $\mathbf{P}$, however, orthogonality of the basis functions is not enough, due to their

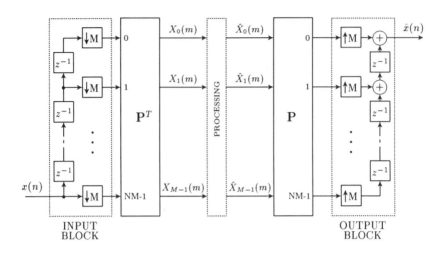

Figure 4.2. Block diagram of a general signal processing system using a lapped transform **P**.

overlapping. There are additional orthogonality constraints among the overlapping portions of the basis functions, which we analyze first in the time domain. Later, in a frequency-domain analysis, we see that all orthogonality constraints correspond to lossless polyphase component matrices.

4.1.1   Time-Domain Analysis

The necessary conditions for perfect signal reconstruction with a lapped transform **P** are easy to obtain when we look at the input and output signals in Fig. 4.1 as infinite vectors, as suggested in [1]. Let us call $\tilde{\mathbf{x}}$ the infinite input vector, $\hat{\tilde{\mathbf{x}}}$ the infinite reconstructed vector, and $\tilde{\mathbf{X}}$ the infinite transform vector in Fig. 4.1. Then, we have

$$\tilde{\mathbf{X}} = \tilde{\mathbf{P}}^T \tilde{\mathbf{x}} \tag{4.3}$$

and, assuming that the transform coefficients are unaltered before computation of the inverse transform,

$$\hat{\tilde{\mathbf{x}}} = \tilde{\mathbf{P}} \tilde{\mathbf{X}} \tag{4.4}$$

where the infinite matrix $\tilde{\mathbf{P}}$ is given by

$$\tilde{\mathbf{P}} = \begin{bmatrix} \ddots & & & & & \mathbf{0} \\ & \mathbf{P} & & & & \\ & & \mathbf{P} & & & \\ & & & \mathbf{P} & & \\ & & & & \mathbf{P} & \\ \mathbf{0} & & & & & \ddots \end{bmatrix}$$

In the equation above, each block $\mathbf{P}$ is positioned M rows below the $\mathbf{P}$ block to its left, and thus M rows above the $\mathbf{P}$ block to its right. Without loss of generality, let us assume that $L = NM$, where N determines the number of overlapping signal blocks that are used for each LT block. Thus, the last column of a $\mathbf{P}$ block and the first column of the next $\mathbf{P}$ block may have up to $(N-1)M$ nonzero elements in the same rows. If $N = 1$, then there is no overlap, the matrix $\tilde{\mathbf{P}}$ becomes block diagonal, and the lapped transform $\mathbf{P}$ becomes a square transform. In this sense, traditional block transforms can be viewed as special cases of lapped transforms with $N = 1$.

Combining (4.3) and (4.4) we get

$$\hat{\tilde{\mathbf{x}}} = \tilde{\mathbf{P}}\tilde{\mathbf{P}}^T\tilde{\mathbf{x}}$$

Perfect reconstruction means $\hat{\tilde{\mathbf{x}}} = \tilde{\mathbf{x}}$, which is achieved if and only if

$$\tilde{\mathbf{P}}\tilde{\mathbf{P}}^T = \mathbf{I} \quad \Leftrightarrow \quad \tilde{\mathbf{P}}^T\tilde{\mathbf{P}} = \mathbf{I}$$

The equations above mean not only that the basis functions in $\mathbf{P}$ must be orthogonal among themselves, but also that they must be orthogonal to the overlapping portions of the basis functions corresponding to neighboring blocks. Thus, they are equivalent to

$$\mathbf{P}^T\mathbf{W}^m\mathbf{P} = \delta(m)\,\mathbf{I}, \quad m = 0, 1, \ldots, N-1 \tag{4.5}$$

where $\mathbf{W}$ is the one-block shift matrix, defined by

$$\mathbf{W} \equiv \begin{pmatrix} \mathbf{0} & \mathbf{I} \\ \mathbf{0} & \mathbf{0} \end{pmatrix} \tag{4.6}$$

where the identity matrix above is of order $(N-1)M$.

The restrictions in (4.5) are the necessary and sufficient conditions for a $NM \times M$ matrix to be a lapped transform. For $m = 0$, they require the orthogonality of the basis functions (the columns of $\mathbf{P}$), whereas for $m > 0$ they are known as the *orthogonality of the tails* conditions [1, 2].

4.1.2   Connection with Filter Banks

Comparing the basic LT block diagram of Fig. 4.2 with the multirate filter bank of Fig. 3.7, it is clear that the former is a special case of the latter. Hence, a lapped transform is a filter bank in which the impulse responses of the synthesis filters are the LT basis functions, and the impulse responses of the analysis filters are the time-reversed basis functions, that is,

$$f_k(n) = p_{nk}, \quad k = 0, 1, \ldots, M - 1, \quad n = 0, 1, \ldots, NM - 1 \tag{4.7}$$

and

$$h_k(n) = f_k(NM - 1 - n) = p_{NM-1-n,k} \tag{4.8}$$

where p_{nk} is the element in the nth row and kth column of $\mathbf{P}$. In the right-hand side of (4.8) we have used a comma to make clear the separation between the two indices of p. Thus, both p_{nk} and $p_{n,k}$ denote the nth sample of the kth basis function.

The transfer function of the kth analysis filter is given by

$$H_k(z) = \sum_{r=0}^{NM-1} h_k(r)\, z^{-r} = \sum_{r=0}^{NM-1} p_{NM-1-r,k}\, z^{-r} \tag{4.9}$$

and the transfer function of the kth synthesis filter is

$$F_k(z) = \sum_{r=0}^{NM-1} f_k(r)\, z^{-r} = \sum_{r=0}^{NM-1} p_{rk}\, z^{-r} \tag{4.10}$$

We recall from Section 3.4 that the transfer functions above can be written in terms of their polyphase components, in the forms

$$H_k(z) = \sum_{r=0}^{M-1} z^{-r}\, E_{kr}(z^M) \tag{4.11}$$

and

$$F_k(z) = \sum_{r=0}^{M-1} z^{-(NM-1-r)}\, \tilde{E}_{rk}(z^M) \tag{4.12}$$

Comparing (4.9) and (4.10) to (4.11) and (4.12), we see that the rth polyphase component of $H_k(z)$ is given by

$$E_{kr}(z) = \sum_{m=0}^{N-1} p_{(N-m)M-1-r,k} \, z^{-m}$$

Thus, the polyphase component matrix $\mathbf{E}(z)$ is given by

$$\mathbf{E}(z) = \sum_{l=0}^{N-1} z^{-(N-1-l)} \mathbf{P}_l^T \, \mathbf{J} \tag{4.13}$$

where $\mathbf{P}_l$ is the lth square block of $\mathbf{P}$, defined by

$$\mathbf{P} \equiv [\mathbf{P}_0^T \ \mathbf{P}_1^T \ \cdots \ \mathbf{P}_{N-1}^T]$$

and $\mathbf{J}$ is the counter-identity matrix [3]

$$\mathbf{J} = \begin{pmatrix} 0 & \cdots & & 0 & 1 \\ 0 & \cdots & 0 & 1 & 0 \\ \vdots & & & & \vdots \\ 1 & 0 & & \cdots & 0 \end{pmatrix} \tag{4.14}$$

With $\mathbf{E}(z)$ given by (4.13), we have

$$\mathbf{E}(z)\tilde{\mathbf{E}}(z) = \sum_{n=0}^{N-1} \sum_{m=0}^{N-1} z^{(n-m)} \mathbf{P}_n^T \mathbf{P}_m \tag{4.15}$$

where we recall from Chapter 3 that $\tilde{\mathbf{E}}(z) \equiv \mathbf{E}^T(z^{-1})$ [4]. The orthogonality conditions in (4.5) are equivalent to

$$\sum_{m=0}^{N-1-l} \mathbf{P}_m^T \mathbf{P}_{m+l} = \delta(l) \, \mathbf{I} \tag{4.16}$$

Replacing the equation above into (4.15), we obtain

$$\mathbf{E}(z)\tilde{\mathbf{E}}(z) = \mathbf{I}$$

Thus, the orthogonality conditions that define lapped transforms in (4.5) imply that the LT filter bank has a lossless polyphase component matrix. Note that, since $\mathbf{E}(z)\tilde{\mathbf{E}}(z) = \mathbf{I}$ implies $\tilde{\mathbf{E}}(z)\mathbf{E}(z) = \mathbf{I}$, (4.16) can also be written in the form

$$\sum_{m=0}^{N-1-l} \mathbf{P}_m \mathbf{P}_{m+l}^T = \delta(l)\,\mathbf{I} \tag{4.17}$$

which is an equivalent necessary and sufficient PR condition.

We have seen in the development above that a lapped transform corresponds to a perfect reconstruction filter bank with FIR filters and a lossless polyphase component matrix. In fact, the connection is stronger because we can show that any uniform PR FIR filter bank is a lapped transform. In order to see that, let us assume that we have a PR filter bank with a lossless polyphase component matrix, i.e., $\mathbf{E}(z)\tilde{\mathbf{E}}(z) = \tilde{\mathbf{E}}(z)\mathbf{E}(z) = \mathbf{I}$. Then, as we saw in Chapter 3, the corresponding aliasing cancellation matrix $\mathbf{H}(z)$ satisfies

$$\tilde{\mathbf{H}}(z)\,\mathbf{H}(z) = \mathbf{H}(z)\,\tilde{\mathbf{H}}(z) = M\,\mathbf{I}$$

and so we can write

$$\sum_{r=0}^{M-1} H_k(z^{-1}W_M^r)\,H_l(zW_M^{-r}) = M\delta(k-l) \tag{4.18}$$

where, as usual, $W_M \equiv e^{-j2\pi/M}$.

Let us define $U_{kl}(z)$ by

$$U_{kl}(z) \equiv H_k(z)\,H_l(z^{-1})$$

which is a generalization of the definition in (3.50). In the time domain, this corresponds to

$$u_k(n) = h_k(n) * h_l(-n)$$

and we can rewrite (4.18) as

$$\sum_{r=0}^{M-1} U_{kl}(z^{-1}W_M^r) = M\delta(k-l)$$

or, in the time domain,

$$u_{kl}(rM) = \delta(r)\,\delta(k - l)$$

which is equivalent to

$$h_k(n) * h_l(-n)\ \Big|_{n=rM} = \delta(r)\,\delta(k - l) \tag{4.19}$$

We recall that $h_k(n) = p_{NM-1-n,k}$, and so the result in (4.19) is identical to the orthogonality conditions in (4.5). Therefore, we have demonstrated that a PR filter bank with a lossless polyphase component matrix is a lapped transform.

It is interesting that the two approaches that were independently followed for the development of PR filter banks, namely lapped transforms (with time-domain orthogonality conditions) and lossless polyphase component matrices (with frequency-domain orthogonality conditions) led to precisely the same class of solutions.

The fundamental problems that must be addressed in order to put lapped transforms to work in practice are:

- How do we design an LT matrix $\mathbf{P}$ such that the orthogonality conditions in (4.5) are satisfied?

- Within the space of valid LT matrices, which are the good ones, in terms of coding and filtering performance?

- Within the space of LTs with good performance, which ones lead to fast computational algorithms?

Fortunately, it is possible to find good LTs with fast algorithms. In fact, one of the major purposes of this book is to present such LTs.

In the following sections we discuss in detail the design techniques and the fast algorithms for the *lapped orthogonal transform*. As we have mentioned before, the LOT historically stands for the lapped transform with $L = 2M$. LTs with longer basis functions are discussed in Chapter 5.

4.2   The Lapped Orthogonal Transform

The LOT was introduced in [1]. At that time, only basis functions with lengths $L = 2M$ were considered, and the intended application was image data compression. Thus, the LOTs presented in [1] were designed in an attempt to maximize the transform coding gain G_{TC} [see Section 2.3].

4.2.1   Recursive LOT Optimization

We recall from Section 2.3 that the covariance of the transformed block $\mathbf{X}$ in (4.1) is given by

$$\mathbf{R_{XX}} = \mathbf{P}^T \, \mathbf{R_{xx}} \, \mathbf{P}$$

Furthermore, we saw that the maximum transform coding gain G_{TC} is obtained when the geometric mean of the diagonal elements of $\mathbf{R_{XX}}$ (which are the variances of the transformed coefficients) is minimized.

In order to maximize G_{TC}, it was suggested in [1] the recursive optimization of the LOT basis functions. The procedure starts with the first basis function $[\mathbf{P}]_{\cdot 0}$ (the first column of $\mathbf{P}$), which is optimized for the maximum variance of $[\mathbf{X}]_0$, under the constraint of the orthogonality of the tails condition in (4.5). Given the optimum $[\mathbf{P}]_{\cdot 0}$, then $[\mathbf{P}]_{\cdot 1}$ is optimized for the maximum variance of $[\mathbf{X}]_1$, and so on, until all basis functions are obtained. The algorithm goes as follows:

Recursive LOT Design Algorithm

Step 1: Set the basis function index k to zero, and find the column vector $[\mathbf{P}]_{\cdot 0}$ that solves the following optimization problem:

$$\begin{aligned} \text{maximize} \quad & [\mathbf{P}]_{\cdot 0}^T \, \mathbf{R_{xx}} \, [\mathbf{P}]_{\cdot 0} \\ \text{subject to} \quad & [\mathbf{P}]_{\cdot 0}^T \, [\mathbf{P}]_{\cdot 0} = 1 \\ \text{and} \quad & [\mathbf{P}]_{\cdot 0}^T \mathbf{W} [\mathbf{P}]_{\cdot 0} = 0 \end{aligned}$$

where $\mathbf{W}$ is the M-sample shift matrix defined in (1.28). The optimum $[\mathbf{P}]_{\cdot 0}$ is the first LOT basis function.

Step 2: Set $k \leftarrow k + 1$, and solve

$$\text{maximize} \quad [\mathbf{P}]_{.k}^T \mathbf{R_{xx}} [\mathbf{P}]_{.k}$$
$$\text{subject to} \quad [\mathbf{P}]_{.k}^T [\mathbf{P}]_{.k} = 1,$$
$$[\mathbf{P}]_{.l}^T [\mathbf{P}]_{.k} = 0, \quad l = 0, 1, \ldots, k - 1,$$
$$\text{and} \quad [\mathbf{P}]_{.l}^T \mathbf{W} [\mathbf{P}]_{.k} = 0, \quad l = 0, 1, \ldots, k$$

The solution is the kth LOT basis function.

Step 3: If $k = M - 1$, stop; otherwise go back to Step 2.

The LOT that is obtained from the algorithm above depends on the assumed signal spectrum, because it uses the signal covariance matrix $\mathbf{R_{xx}}$. In [1], a first-order Markov model with an intersample correlation coefficient of $\rho = 0.95$ was used, since it is a good model for images [5]. As noted in [2], however, the algorithm is relatively insensitive to the assumed signal spectrum.

In order to solve the optimization problems in Steps 1 and 2, the augmented Lagrangian method [6] was used in [1]. Although closed-form equations for the computation of the gradient and the Hessian of the objective functions can be derived [1], convergence of the augmented Lagrangian method is not guaranteed. In fact, as k increases in Step 2, the nonlinearity of the constraints is so strong that the optimization may converge to a local minimum; this actually happened in [1].

The first four LOT basis functions, obtained in [1] for $M = 8$ and with $\mathbf{R_{xx}}$ corresponding to a first order Markov model with $\rho = 0.95$, are shown in Fig. 4.3. Note that the stopband attenuations of the low order LOT basis functions are significantly better than those of the DCT [see Fig. 3.29]. The response of the last basis function though, is actually worse than that of the corresponding DCT basis function. This indicates that the optimization method in Step 2 did not converge to the true optimum.

As we can see in Fig. 4.3, the LOT basis functions have alternately even and odd symmetry, with basis number zero being evenly symmetric. In fact, these symmetries were forced in the recursive LOT design algorithm described in [1], for two reasons: first, the number of unknown variables in Steps 1 and 2 would be halved; second, the efficient block transforms such as the KLT or the DCT have such symmetries. However, there is an even stronger reason for such symmetries. As we have already mentioned in Section 1.2, eigenvectors of symmetric Toeplitz matrices must have this symmetry [7, 8], and the optimization problems in Steps 1 and 2 can be viewed as restricted eigenvector problems.

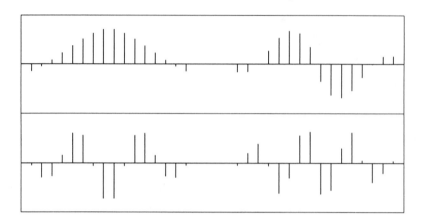

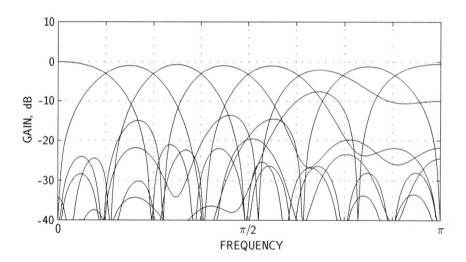

Figure 4.3. Top: the first four basis functions of the LOT designed with the recursive algorithm; bottom: magnitude frequency responses of all basis functions.

We should note that the recursive LOT design algorithm could be used to obtain LOTs for $N > 2$, i.e., $L > 2M$. Then, the more general orthogonality condition of (4.5) would have to be used. That would increase the number of nonlinear constraints in Steps 1 and 2, and therefore the convergence problems would be amplified.

Even if the algorit m were finely tuned in order to minimize the problems of convergence to local minima, it would lead to an LOT matrix $\mathbf{P}$ without any special structure. It would be unlikely, therefore, that a simple factorization for $\mathbf{P}$ could be found. Then, such an optimal LOT would have no fast algorithm, and hence it would be of limited practical usefulness.

4.2.2 Quasioptimal LOTs

In order to make the LOT design more robust, it would be interesting if we could avoid the nonlinear optimization steps. One way to approach this goal, as suggested in [2, Chapter 5], is to work within a subspace of all possible LOT matrices. An optimal LOT within such a subspace would not necessarily be globally optimal, but it might be easier to design and have a fast algorithm.

A basic property of LOT matrices is that, if $\mathbf{P_o}$ is a valid LOT, we can generate a family of LOTs from it in the form

$$\mathbf{P} = \mathbf{P_o} \mathbf{Z} \tag{4.20}$$

where $\mathbf{Z}$ is any orthogonal matrix of order M, i.e., $\mathbf{Z}^T \mathbf{Z} = \mathbf{I}$. This can be verified simply by substituting (4.20) into (4.5). A possible choice for $\mathbf{P_o}$ is [2, 3]

$$\mathbf{P_o} = \frac{1}{2} \begin{bmatrix} \mathbf{D_e} - \mathbf{D_o} & \mathbf{D_e} - \mathbf{D_o} \\ \mathbf{J}(\mathbf{D_e} - \mathbf{D_o}) & -\mathbf{J}(\mathbf{D_e} - \mathbf{D_o}) \end{bmatrix} \tag{4.21}$$

where $\mathbf{D_e}$ and $\mathbf{D_o}$ are the $M \times M/2$ matrices containing the even and odd DCT functions of length M, respectively. It is easy to verify that (4.5) is satisfied for $\mathbf{P_o}$ defined above. There were two reasons for this choice: first, because the first $M/2$ basis functions approximate closely the evenly symmetric ones obtained in [1]; second, because such a matrix $\mathbf{P_o}$ is the key to the fast LOT algorithm to be discussed in the next section.

For a given $\mathbf{P_o}$ and any $\mathbf{Z}$, the covariance matrix of the LOT coefficients is

$$\mathbf{R_{xx}} = \mathbf{Z}^T \mathbf{R_o} \mathbf{Z}$$

where

$$\mathbf{R_o} \equiv \mathbf{P}_o^T \, \mathbf{R_{xx}} \, \mathbf{P}_o$$

The matrix $\mathbf{Z}$ that would maximize the coding gain is the one whose columns are the eigenvectors of $\mathbf{R_o}$, as in the case of the KLT discussed in Section 1.2. Then, we have the following design procedure.

Quasioptimal LOT Design Algorithm
Maximum Coding Gain Version

Step 1: Define a valid LOT matrix $\mathbf{P_o}$ as in (4.21), that is,

$$\mathbf{P_o} = \frac{1}{2} \left[\begin{array}{cc} \mathbf{D_e} - \mathbf{D_o} & \mathbf{D_e} - \mathbf{D_o} \\ \mathbf{J}(\mathbf{D_e} - \mathbf{D_o}) & -\mathbf{J}(\mathbf{D_e} - \mathbf{D_o}) \end{array} \right]$$

This choice of $\mathbf{P_o}$ leads to evenly-symmetric basis functions (the first $M/2$ columns) that are smooth, but it also leads to oddly-symmetric functions (the last $M/2$ columns) with sharp transitions in the middle of the block. This is why we need the following steps.

Step 2: For a given signal covariance matrix $\mathbf{R_{xx}}$, define a new covariance matrix $\mathbf{R_o}$ by

$$\mathbf{R_o} \equiv \mathbf{P}_o^T \, \mathbf{R_{xx}} \, \mathbf{P}_o$$

Step 3: Compute the eigenvectors of $\mathbf{R_o}$. Define $\mathbf{Z}$ as the matrix whose columns are the eigenvectors of $\mathbf{R_o}$. Then the quasioptimal LOT matrix $\mathbf{P}$ is given by

$$\mathbf{P} = \mathbf{P_o} \, \mathbf{Z}$$

The first four LOT basis functions obtained in [2] with the algorithm above, for $M = 8$, are shown in Fig. 4.4. Again, $\mathbf{R_{xx}}$ was that of a first order Markov model with $\rho = 0.95$. Besides the better frequency responses of the quasioptimal LOT, compared to those in Fig. 4.3, it is interesting to note that the corresponding coding gain was 9.24 dB, whereas the LOT in Fig. 4.3 has a coding gain of 9.17 dB. Thus, it is clear that the recursive LOT design algorithm had convergence problems. Furthermore, these numbers show that the coding gain of the quasioptimal LOT is probably quite close to the maximum that could be achieved.

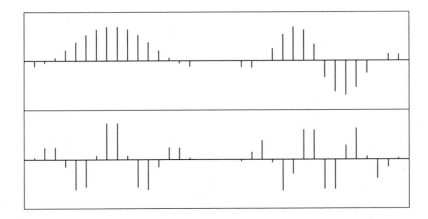

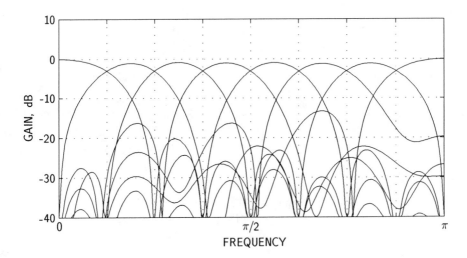

Figure 4.4. Top: the first four basis functions of the quasioptimal LOT obtained as the solution to an eigenvector problem; bottom: magnitude frequency responses of all basis functions.

Another design technique for a quasioptimal LOT was suggested in [9], with the goal of improving the frequency responses of the LOT basis functions. The technique was based on the observation that the evenly symmetrical LOT functions in the first $M/2$ columns in (4.21) are already good bandpass filters. Using the simple low-pass to high-pass transformation $h(n) \rightarrow (-1)^n h(n)$, we redefine the last $M/2$ columns in (4.21), in the form

$$\mathbf{P_s} = \frac{1}{2} \begin{bmatrix} \mathbf{D_e} - \mathbf{D_o} & \mathbf{U}(\mathbf{D_e} - \mathbf{D_o})\mathbf{J} \\ \mathbf{J}(\mathbf{D_e} - \mathbf{D_o}) & \mathbf{UJ}(\mathbf{D_e} - \mathbf{D_o})\mathbf{J} \end{bmatrix} \qquad (4.22)$$

where $\mathbf{U} \equiv \mathrm{diag}\{1, -1, 1, \ldots, 1, -1\}$. Note that pre-multiplication by $\mathbf{U}$ changes the symmetry from even to odd, and post-multiplication by $\mathbf{J}$ serves to put the last $M/2$ columns in ascending order of center frequency of their bandpass responses.

The matrix $\mathbf{P_s}$ above does not lead to perfect reconstruction, because $\mathbf{P_s}^T \mathbf{W} \mathbf{P_s} \neq \mathbf{0}$. However, the PR error is relatively small, so that $\mathbf{P_s}$ could be used as an LT in low bit-rate coding applications. This choice of $\mathbf{P_s}$ has been known as the pseudolapped orthogonal transform [10].

In order to map the matrix $\mathbf{P_s}$ of (4.22) into a PR LOT, consider the matrix $\mathbf{A}$ defined by

$$\mathbf{A} \equiv \mathbf{P_o}^T \mathbf{P_s}$$

the orthogonal matrix $\mathbf{Z}$ that is closest to $\mathbf{A}$ can be obtained by means of the orthogonal factor of the QR factorization [11] of $\mathbf{A}$. Then, a valid LOT that approximates $\mathbf{P_s}$ is given by $\mathbf{P_o}\mathbf{Z}$. Summarizing, the steps of this design procedure are the following.

Quasioptimal LOT Design Algorithm
Version with Improved Frequency Responses

Step 1: Define a valid LOT matrix $\mathbf{P_o}$ as in (4.21), that is,

$$\mathbf{P_o} = \frac{1}{2} \begin{bmatrix} \mathbf{D_e} - \mathbf{D_o} & \mathbf{D_e} - \mathbf{D_o} \\ \mathbf{J}(\mathbf{D_e} - \mathbf{D_o}) & -\mathbf{J}(\mathbf{D_e} - \mathbf{D_o}) \end{bmatrix}$$

This is the same matrix $\mathbf{P_o}$ that was used in the quasioptimal LOT with maximum coding gain.

Step 2: Define a matrix $\mathbf{P_s}$ with good bandpass filters as in (4.22), that is,

$$\mathbf{P_s} = \frac{1}{2} \begin{bmatrix} \mathbf{D_e} - \mathbf{D_o} & \mathbf{U}(\mathbf{D_e} - \mathbf{D_o})\mathbf{J} \\ \mathbf{J}(\mathbf{D_e} - \mathbf{D_o}) & \mathbf{UJ}(\mathbf{D_e} - \mathbf{D_o})\mathbf{J} \end{bmatrix}$$

The oddly-symmetric basis functions (the last $M/2$ columns of $\mathbf{P_s}$) defined above correspond to a low-pass to high-pass transformation on the evenly-symmetric basis functions.

Step 3: Compute a matrix $\mathbf{A}$ by

$$\mathbf{A} = \mathbf{P_o}^T \mathbf{P_s}$$

Step 4: Obtain the QR factorization [11] of $\mathbf{A}$, i.e., find the orthogonal matrix $\mathbf{Q}$ and the upper triangular matrix $\mathbf{R}$ such that

$$\mathbf{A} = \mathbf{QR}$$

and set $\mathbf{Z} = \mathbf{Q}$. Then the quasioptimal LOT matrix $\mathbf{P}$ is given by

$$\mathbf{P} = \mathbf{P_o Z}$$

An advantage of this quasioptimal LOT design algorithm is that no knowledge of the input signal spectrum is necessary. The first four LOT basis functions obtained in [9] with the algorithm above are shown in Fig. 4.5, for $M = 8$. The frequency responses of this QR-based quasioptimal LOT are also shown in Fig. 4.5. It is clear that they have the best stopband attenuations of all the LOTs described so far, about 21 dB in the worst case. For the AR(1) signal model with $\rho = 0.95$, the coding gain of the QR-based LOT is 9.19 dB, that is only 0.05 dB below that of the quasioptimal LOT designed for maximum coding gain.

Therefore, it seems that the QR-based LOT is the most appropriate for practical applications, since it improves the worst-case stopband attenuations of the other LOTs by more than 6 dB. The price paid for that is a reduction of only 0.05 dB in the coding gain for an AR(1) signal with $\rho = 0.95$.

None of the LOTs described in this section have fast algorithms, because their $\mathbf{P}$ matrices cannot be easily factored. In the next section we discuss the fast versions of the quasioptimal LOTs.

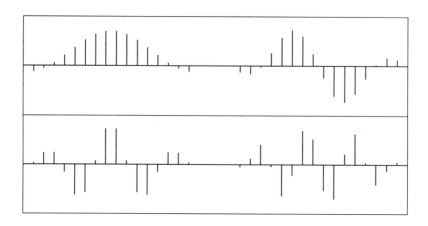

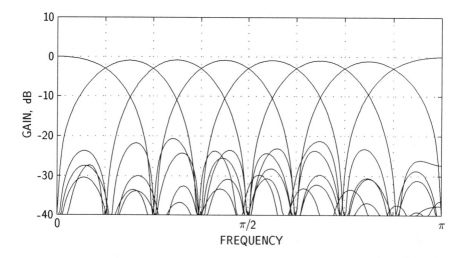

Figure 4.5. Top: the first four basis functions of the quasioptimal LOT obtained from a QR factorization; bottom: magnitude frequency responses of all basis functions.

4.3 Fast Algorithms for the LOT

In order to apply the LOT to work in practical applications, a fast LOT transformation algorithm is necessary. We had precisely that in mind when we built the $\mathbf{P_o}$ matrix in (4.21) from the DCT basis functions because it is clear from (4.21) that $\mathbf{P_o}$ can be efficiently implemented. However, the quasioptimal LOTs have an extra factor, the orthogonal matrix $\mathbf{Z}$. If $\mathbf{Z}$ had no simple factorization, the LOT would still be computationally unattractive.

4.3.1 Structure of the Fast LOT

Fortunately, it turns out that the two quasioptimal LOT design methods described in the previous section lead to orthogonal factors $\mathbf{Z}$ that can be approximated by

$$\mathbf{Z} \simeq \begin{bmatrix} \mathbf{I} & \mathbf{0} \\ \mathbf{0} & \tilde{\mathbf{Z}} \end{bmatrix} \tag{4.23}$$

where $\tilde{\mathbf{Z}}$ is an orthogonal matrix of order $M/2$. It is important to note that such an approximation means that the first $M/2$ basis functions in $\mathbf{P_o}$ (the ones with even symmetry) are close to be optimal for coding and filtering. In fact, it was shown in [2] that they are asymptotically optimal for an AR(1) signal with $\rho \to 1$.

The matrix $\tilde{\mathbf{Z}}$ affects only the last $M/2$ basis functions in $\mathbf{P}$, the ones with odd symmetry. For $M \leq 16$, it was shown in [2, 9] that the approximation in (4.23) is still good enough when $\tilde{\mathbf{Z}}$ is constrained to be a cascade of $M/2 - 1$ plane rotations, in the form

$$\tilde{\mathbf{Z}} = \mathbf{T_0 T_1} \cdots \mathbf{T}_{M/2-2} \tag{4.24}$$

where each plane rotation is defined by

$$\mathbf{T}_i = \begin{bmatrix} \mathbf{I} & \mathbf{0} & \mathbf{0} \\ \mathbf{0} & \mathbf{Y}(\theta_i) & \mathbf{0} \\ \mathbf{0} & \mathbf{0} & \mathbf{I} \end{bmatrix} \tag{4.25}$$

The first identity factor in the above equation is of order i, that is, the top left element of $\mathbf{Y}(\theta_i)$ is in position (i, i) of the $M/2 \times M/2$ matrix $\mathbf{T}_i$ (indices start at zero).

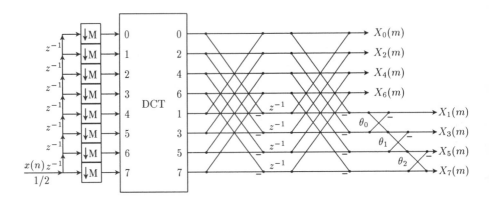

Figure 4.6. Flowgraph of the fast LOT, type I, direct transform, for $M = 8$.

The matrix $\mathbf{Y}(\theta_i)$ is a 2×2 butterfly,

$$\mathbf{Y}(\theta_i) = \left[\begin{array}{cc} \cos\theta_i & \sin\theta_i \\ -\sin\theta_i & \cos\theta_i \end{array} \right] \tag{4.26}$$

where θ_i is the rotation angle. It is clear that $\mathbf{Y}(\theta_i)$ is an orthogonal matrix.

The flowgraph of the fast direct LOT is shown in Fig. 4.6, and the corresponding inverse LOT flowgraph is shown in Fig. 4.7. Note that only $3M/2 - 1$ delay units are necessary in each, compared to the $2M - 1$ delay units that would be necessary in the simple configuration of Fig. 4.2. We refer to these fast LOTs as the type-I fast LOT, to distinguish from the fast LOT that we consider in the next section.

Suppose we already have a DCT-based signal processing system, such as an adaptive image coder, for example. It is clear from Figs. 4.6 and 4.7 that modifying such a system to operate with the LOT would be quite easy. The existing DCT/IDCT hardware or software would be kept, and we would have to add to the system the software or hardware to implement the ± 1 butterflies and the $\mathbf{Y}(\theta_i)$ butterflies.

Certainly, using the LOT is computationally more expensive than the DCT. Assuming that each nontrivial butterfly takes three multiplications and three additions, a DCT versus LOT comparison for $M = 8$ would give us the following result:

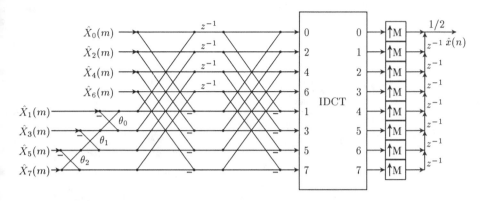

Figure 4.7. Flowgraph of the fast LOT, type I, inverse transform, for $M = 8$.

TRANSFORM	NO. OF MULTIPLICATIONS	NO. OF ADDITIONS	TOTAL
DCT	13	29	42
LOT	22	54	76

Thus, we see that there is an overhead of about 80 percent in the total number of arithmetic operations for the LOT, compared to the DCT. This is a relatively small overhead in many practical applications. For video coding with a sampling rate of 10 MHz, for example, real-time computation of a length-8 transform must be formed in 800 ns. For the complexities listed above, the design of integrated circuits for real-time computation of either the DCT or the LOT, for video applications, is quite within the range of current CMOS technology.

Using the fast LOT implementation, we can approximate closely the basis functions of the quasioptimal LOTs presented in the previous section. The angles that approximate the LOT optimized for maximum coding gain are $[\theta_0 \; \theta_1 \; \theta_2] = [0.13\pi \; 0.16\pi \; 0.13\pi]$, and the angles that approximate the QR-based quasioptimal LOT are $[\theta_0 \; \theta_1 \; \theta_2] = [0.145\pi \; 0.17\pi \; 0.16\pi]$. The first four basis functions and the magnitude frequency responses of these two LOTs are shown in Figs. 4.8 and 4.9, respectively.

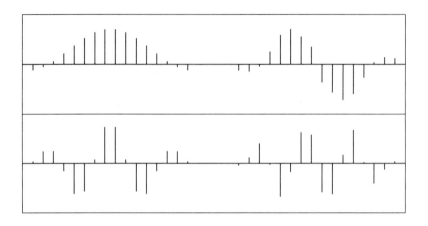

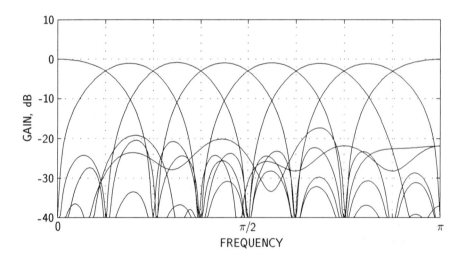

Figure 4.8. Top: the first four basis functions of the type-I fast LOT with maximum coding gain; bottom: magnitude frequency responses of all basis functions.

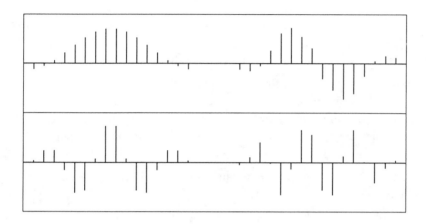

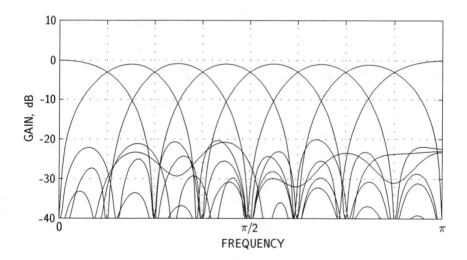

Figure 4.9. Top: the first four basis functions of the type-I fast LOT approximating the QR-based LOT; bottom: magnitude frequency responses of all basis functions.

4.3.2 LOTs of Finite-Length Signals

In image coding applications, a two-dimensional LOT would typically be separable, and so we would need to compute the LOT of the rows and the columns of the image. Each row or column is a finite-length signal, and so the fast LOT structures of Figs. 4.6 and 4.7 cannot be directly used for the entire finite-length signal.

With a regular block transform, we simply require that the number of samples in the signal be a multiple of the block size M. With LOTs, however, we need a slight modification in the computation of the transforms of the first and last blocks. Otherwise, the corresponding LOT basis functions would extend outside the region of support of the signal.

For a finite-length input vector, the infinite matrix $\tilde{\mathbf{P}}$ in (4.3) would be replaced by the finite matrix

$$
\tilde{\mathbf{P}} = \begin{bmatrix} \mathbf{P_a} & & & & \mathbf{0} \\ & \mathbf{P} & & & \\ & & \ddots & & \\ & & & \mathbf{P} & \\ \mathbf{0} & & & & \mathbf{P_b} \end{bmatrix}
$$

The LOT matrices of the first and last blocks, $\mathbf{P_a}$ and $\mathbf{P_b}$, respectively, would generate basis functions of length $3M/2$, because there can be no overlapping outside the signal region of support.

A simple way to define $\mathbf{P_a}$ and $\mathbf{P_b}$, as suggested in [3], is to assume that the signal is evenly reflected at its boundaries. The resulting fast LOT structure is shown in Fig. 4.10, where $\mathbf{H_e}$ is the matrix containing half of the samples of the even DCT functions, that is,

$$
\mathbf{D_e} \equiv \begin{bmatrix} \mathbf{H_e} \\ \mathbf{JH_e} \end{bmatrix}
$$

where, as before, the columns of $\mathbf{D_e}$ are the even DCT functions of length M. The dashed portion of Fig. 4.10 shows the part of the flowgraph that is responsible for the computation of the LOT of an inner block. It is easy to see that it corresponds precisely to a noncausal version of the fast LOT in Fig. 4.6. With the LOT structure of Fig. 4.10, there are no visible additional distortions at the borders of the signal in low- and medium-rate image coding, as shown in [3].

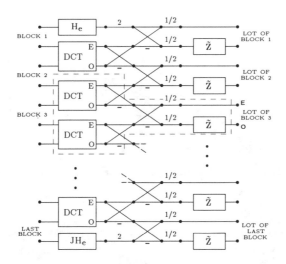

Figure 4.10. Simplified flowgraph of the fast LOT for finite-length signals. The direct LOT is computed from left to right, and the inverse LOT from right to left.

4.4   Fast LOT for $M > 16$

The fast LOT implementations of Figs. 4.6 and 4.7 are valid only for $M \leq 16$, as we have mentioned before. There are applications such as apeech coding [12] and adaptive filtering, where a large number of subbands may be necessary, e.g. $M = 128$ or more. In these applications, the type-I fast LOT cannot be employed.

A more general fast LOT algorithm can be obtained by a different factorization of the matrix $\tilde{\mathbf{Z}}$ in (4.23), suggested in [12], which is

$$\tilde{\mathbf{Z}} = \mathbf{D}_1^T \mathbf{D}_2 \tag{4.27}$$

where $\mathbf{D}_1$ is the matrix whose columns are the basis functions of the DCT of length $M/2$, and $\mathbf{D}_2$ is the type-IV discrete sine transform (DST-IV) matrix, defined in (1.24). Note that the matrix $\mathbf{D}_1$ in (4.27) appears transposed, and so it corresponds to an inverse DCT in the computation of the direct LOT, and to a direct DCT in

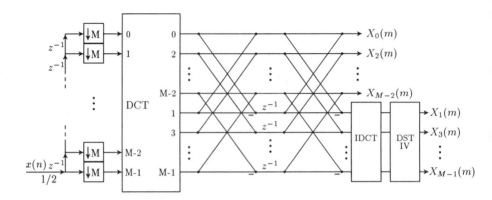

Figure 4.11. Flowgraph of the fast LOT, type II, direct transform. This is a generic version, valid for any M.

the inverse LOT computation. Since both the DCT and the DST-IV have fast algorithms, as we have seen in Chapter 2, the fast LOT based on (4.20), (4.23), and (4.27) is fast-computable.

The flowgraph of this fast LOT, which we will refer to the type-II LOT, is shown in Fig. 4.11, for the direct transform. The corresponding inverse transform flowgraph can be easily obtained by transposing the flowgraph and moving the delay units, as we did in obtaining the inverse LOT in Fig. 4.7 from the direct LOT in Fig. 4.6.

The first four basis functions of the type-II fast LOT are shown in Fig. 4.12, for $M = 8$. The magnitude frequency responses of all basis functions are also shown in Fig. 4.12. We note that the responses of the type-II fast LOT are close to those of the fast LOT approximating the QR-based LOT, which were shown in Fig. 4.9. However, the computational complexity of the type-II LOT is higher: 26 multiplications and 66 additions versus 22 and 54, respectively, for the type-I fast LOT. Thus, for $M \leq 16$ the type-I LOT is more efficient, whereas the type-II LOT must be employed when $M > 16$.

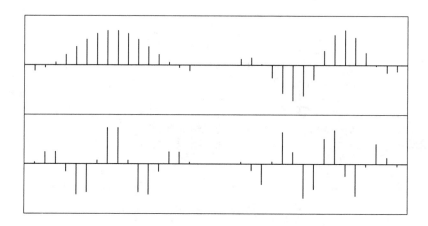

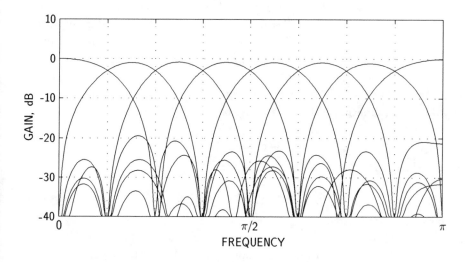

Figure 4.12. Top: the first four basis functions of the type-II fast LOT, for $M = 8$; bottom: magnitude frequency responses of all basis functions.

TRANSFORM	CODING GAIN, dB			MIN. STOPBAND
	AR(1)	MALE SPEECH	FEMALE SPEECH	ATTENUATION, dB
DCT	8.83	7.49	4.27	10
LOT – RECURSIVE	9.17	7.82	4.46	8
LOT – MAX. GAIN	9.24	8.67	4.92	14
LOT – QR	9.20	8.50	5.12	21
LOT – FAST 1	9.20	8.63	4.58	17
LOT – FAST 2	9.19	8.75	4.72	20
TYPE-II LOT	9.22	8.67	5.07	19

Table 4.1. Coding and filtering performance of several LOTs, for $M = 8$.

In general, the overhead associated in the computation of the type-II fast LOT in Fig. 4.11, compared to the DCT, is that associated with the two length-$M/2$ transforms and the ± 1 butterflies in Fig. 4.11. Considering the complexities in Tables 2.1 and 2.2, the number of multiplications for the type-II fast LOT is precisely twice that for the DCT, and the number of additions is slightly over twice the number of additions for the DCT. Thus, the computational overhead in moving from the DCT to the LOT is approximately 100 percent, which in most applications would not be a limiting factor for the use of the LOT.

4.5   Coding Performances

It is interesting to compare the coding performances of the various LOTs that we have discussed in the previous sections. Such a comparison will help us in deciding which LOT is best suited for a particular application. As we have done in Section 2.3, we will consider the coding gain for an AR(1) signal model with $\rho = 0.95$, and for two speech spectra estimated from male and female speech.

Table 4.1 presents a summary of the coding gains of the LOTs. As a standard of comparison, we have also included in the table the gains for the DCT. In the last column, we have included the worst-case stopband attenuation of the basis functions. The rows labeled "LOT – FAST 1" AND "LOT – FAST 2" correspond to the

-0.0684	0.0231	0.0379	0.0469	-0.0716	-0.0273	0.0018	0.0545
-0.0311	0.1444	0.0684	-0.0921	-0.0915	0.0650	0.1477	-0.0239
0.0379	0.1495	-0.2255	0.0231	-0.0317	0.2447	-0.1139	-0.0363
0.1280	-0.0921	-0.0311	0.1495	0.1555	-0.0069	-0.1073	0.1116
0.2255	-0.3699	0.3846	-0.3409	0.3652	-0.3963	0.3504	-0.1964
0.3157	-0.3409	-0.1280	0.4388	0.4335	-0.0788	-0.3825	0.2904
0.3846	0.0469	-0.4220	-0.3699	0.3247	0.3967	-0.0458	-0.4473
0.4220	0.4388	0.3157	0.1444	0.1859	0.3383	0.4272	0.3996
0.4220	0.4388	0.3157	0.1444	-0.1859	-0.3383	-0.4272	-0.3996
0.3846	0.0469	-0.4220	-0.3699	-0.3247	-0.3967	0.0458	0.4473
0.3157	-0.3409	-0.1280	0.4388	-0.4335	0.0788	0.3825	-0.2904
0.2255	-0.3699	0.3846	-0.3409	-0.3652	0.3963	-0.3504	0.1964
0.1280	-0.0921	-0.0311	0.1495	-0.1555	0.0069	0.1073	-0.1116
0.0379	0.1495	-0.2255	0.0231	0.0317	-0.2447	0.1139	0.0363
-0.0311	0.1444	0.0684	-0.0921	0.0915	-0.0650	-0.1477	0.0239
-0.0684	0.0231	0.0379	0.0469	0.0716	0.0273	-0.0018	-0.0545

Table 4.2. Basis functions of the recommended LOT, for $M = 8$. They are generated by the type-I fast LOT with angles $[\theta_0\ \theta_1\ \theta_2] = [0.145\pi\ 0.17\pi\ 0.16\pi]$.

type-I fast LOTs with the butterfly angles that approximate the LOT designed for maximum coding gain and the one designed with the QR factorization, respectively.

We believe that, for $M < 16$ the best compromise among coding gain, filtering performance, and computational complexity, is that of the fast LOT with the angles approximating the LOT designed with the QR factorization. Thus, we present in Table 4.2 the coefficients of this fast LOT, which we suggest to be the standard LOT for all applications.

4.6 Summary

The purpose of this chapter has been to present the basic theory of lapped transforms (LTs), and the details of the design procedures and performances of several versions of the lapped orthogonal transform (LOT). In Section 4.1 we saw that the time-domain extended orthogonality conditions that define LTs are equivalent to the frequency-domain losslessness conditions that define perfect-reconstruction FIR filter banks. Thus, all lapped transforms are PR FIR filter banks with identical analysis and synthesis filters (within time reversal), and viceversa.

The design techniques for the LOT were discussed in Section 4.2. The LOT is a lapped transform whose basis functions have linear phase and lengths equal to $L = 2M$. The type-I fast LOT algorithm, which can be used for $M \leq 16$, was presented in Section 4.3, and the type-II fast LOT, which valid for any M, was presented in Section 4.4.

We saw in Section 4.5 that the best LOT for practical applications is the fast LOT that approximates the one designed with the QR factorization. Thus, we have included in Table 4.2 the coefficients of the basis functions of that LOT, which we recommend for all applications with $M = 8$. For other applications, we recommend the type-II fast LOT.

All the LOTs that we have discussed in this chapter have basis functions with even or odd symmetry. It is easy to verify that symmetrical basis functions cannot satisfy the orthogonality constraints in (4.5) for $M = 2$. Also, our quasioptimal LOTs are based on constructions that assume M even. Therefore, the LOT is defined only for $M \geq 4$.

References

[1] Cassereau, P., *A New Class of Optimal Unitary Transforms for Image Processing*, Master's thesis, Mass. Inst. Tech., Cambridge, MA, May 1985.

[2] Malvar, H. S., *Optimal Pre- and Post-filters in Noisy Sampled-Data Systems*, PhD thesis, Mass. Inst. Tech., Cambridge, MA, Sep. 1986.

[3] Malvar, H. S., and D. H. Staelin, "The LOT: transform coding without blocking effects," *IEEE Trans. Acoust., Speech, Signal Processing*, vol. 37, Apr. 1989, pp. 553–559.

[4] Vaidyanathan, P. P., "Theory and design of M-channel maximally decimated quadrature mirror filters with arbitrary M, having the perfect reconstruction property," *IEEE Trans. Acoust., Speech, Signal Processing*, vol. ASSP-35, Apr. 1987, pp. 476–492.

[5] Clarke, R. J., *Transform Coding of Images*, London: Academic Press, 1985.

[6] Bertsekas, D., *Constrained Optimization and Lagrange Multiplier Methods*, New York: Academic Press, 1982.

[7] Cantoni, A., and P. Butler, "Eigenvalues and eigenvectors of symmetric centrosymmetric matrices," *Lin. Algebra Appl.*, vol. 13, 1976, pp. 275–288.

[8] Makhoul, J., "On the eigenvectors of symmetric Toeplitz matrices," *IEEE Trans. Acoust., Speech, Signal Processing*, vol. ASSP-29, Aug. 1981, pp. 868–872.

[9] Malvar, H. S., "The LOT: a link between block transform coding and multirate filter banks," *IEEE Intl. Symp. Circuits Syst.*, Espoo, Finland, pp. 835–838, June 1988.

[10] Malvar, H. S., "Pseudo lapped orthogonal transform," *Electron. Lett.*, vol. 25, no. 5, March 1989, pp. 312–314.

[11] Dahlquist, G., and A. Björk, *Numerical Methods*, Englewood Cliffs, NJ: Prentice-Hall, 1974.

[12] Malvar, H. S., and R. Duarte, "Transform/subband coding of speech with the lapped orthogonal transform," *IEEE Intl. Symp. Circuits Syst.*, Portland, OR, pp. 1268–1271, May 1989.

Chapter 5

Modulated Lapped Transforms

This chapter describes in detail the theory, design techniques, fast algorithms, and theoretical performances of a family of lapped orthogonal transforms that is generated from sinusoidal modulations of a low-pass prototype. As we saw Section 3.3, almost-perfect reconstruction filter banks with a large number of subbands could be obtained from this sinusoidal modulation concept; an example is the pseudo-QMF filter bank. One of the main goals of this chapter is to show that perfect reconstruction can actually be achieved, with appropriate choices of the phases of the modulation sinusoids and the low-pass prototype $h(n)$.

In Section 5.1 we review the modulated lapped transform (MLT), which is similar to the LOT discussed in Chapter 4, in the sense that the basis functions have lengths equal to $L = 2M$, where M is the number of subbands. Until quite recently, it was thought that perfect reconstruction with longer basis functions was not possible, but this is not true. In Section 5.2 we see that perfect reconstruction in a modulated filter bank is possible with basis functions of arbitrary length. When $L > 2M$, we refer to the transform as an extended lapped transform (ELT). With the ELT, the system designer can achieve a better trade-off between performance and computational complexity, for the application at hand.

As it was the case with the LOT, there are certain degrees of freedom in defining the basis functions of the MLT and ELT, for a given number of subbands M and filter length L. We can choose them optimally according to a number of design criteria. The one that seems to be more appropriate in practice is to optimize the filtering performance, as we will discuss in detail in Section 5.3.

The main advantage of the MLT-ELT filter banks is that they can be computed efficiently, with computational complexities that are in the same order of magnitude as those of standard block transforms. We will present the details of the fast MLT-ELT algorithms in Section 5.4.

The theoretical performances of the MLT and ELT for coding applications will be discussed in Section 5.5. The MLT and ELT can achieve significant improvements in the coding gain G_{TC}, when compared to standard block transforms. Another major advantage of the MLT and ELT is the absence of blocking effects.

5.1   The MLT

When the length L of the filters is constrained to be equal to twice the number of subbands, i.e., $L = 2M$, the filter banks based on *time-domain aliasing cancellation* (TDAC) [1, 2] can achieve perfect reconstruction. There are two kinds of TDAC filter banks. With even channel stacking, there are $M + 1$ subbands, and the first and last subbands have half the bandwidth of the others [1]. With odd channel stacking, the TDAC filter bank has M subbands of equal bandwidth [2], and the filter responses can be put in the modulated form of (3.22).

It was later recognized [3] that the oddly-stacked TDAC filter bank actually corresponds to a lapped transform, since all perfect-reconstruction filter banks with identical analysis and synthesis filters (within time reversal) are lapped transforms, as we saw in Chapter 4 (the evenly-stacked TDAC filter bank is not a lapped transform). Thus, we refer to the oddly-stacked TDAC filter banks as the modulated lapped transform (MLT). Note that the word orthogonal has been dropped, but the MLT satisfies the same orthogonality conditions as the LOT.

The basis functions of the MLT are the impulse responses of the corresponding synthesis filter bank, and they are defined, as in (3.22), by

$$p_{nk} = f_k(n) = h(n) \cos \left[\left(k + \frac{1}{2} \right) \left(n - \frac{L-1}{2} \right) \frac{\pi}{M} + \phi_k \right]$$

for $k = 0, 1, \ldots, M - 1$, and $n = 0, 1, \ldots, L - 1$. As usual, M is the number of subbands (the number of transform coefficients), and L is the length of the filters. Assuming that $L = NM$, the choice of the phases of the modulation functions that leads to perfect reconstruction is

$$\phi_k = \left(k + \frac{1}{2} \right) (N + 1) \frac{\pi}{2}$$

In fact, the phases in [2] are not identical to the ones above, but differ by π in some of the angles. Such a difference is irrelevant, because it does not affect the shapes of the basis functions, but only their relative signs.

With the phases ϕ_k above, the MLT basis functions can be written in the form

$$p_{nk} = h(n)\sqrt{\frac{2}{M}}\cos\left[\left(n + \frac{M+1}{2}\right)\left(k + \frac{1}{2}\right)\frac{\pi}{M}\right] \tag{5.1}$$

for $k = 0, 1, \ldots, M - 1$, and $n = 0, 1, \ldots, 2M - 1$. The factor $\sqrt{2/M}$ was introduced in order to normalize the basis functions. In [2] it was shown that a filter bank defined as in the equation above has perfect reconstruction when the following conditions on the low-pass prototype $h(n)$ (also refereed as the window) are met:

$$h(L - 1 - n) = h(n) \tag{5.2}$$

which imposes even symmetry of the window, and

$$h^2(n) + h^2(n + M) = 1 \tag{5.3}$$

The proof of the conditions above will be given in the next section, as a special case of the PR conditions for the ELT.

A possible choice for the window, for example [4], is

$$h(n) = \pm\sin\left[\frac{n\pi}{2(M - 1)}\right]$$

Such a window satisfies the perfect reconstruction condition, but it leads to a low-pass subband filter $f_0(n)$ that is not polyphase normalized.

The *polyphase normalization* [5] property states that a dc input signal should be perfectly reconstructed with only the low-pass subband, i.e., if

$$x(n) \equiv 1$$

then

$$|X_k(m)| = \begin{cases} 1/\sqrt{M}, & k = 0 \\ 0, & k = 1, 2, \ldots, M - 1 \end{cases}$$

This is a desirable property in image coding, for example, where lack of polyphase normalization usually leads to visible artifacts in the reconstructed image [5].

In [5] it was shown that the unique window that satisfies both polyphase normalization and perfect reconstruction is

$$h(n) = \pm \sin\left[\left(n + \frac{1}{2}\right)\frac{\pi}{2M}\right] \tag{5.4}$$

The choice of the sign in the equation above is irrelevant, since it will only determine the sign of the low-pass subband. If we want $X_0(m)$ to be positive with a positive dc input, the sign should be a minus.

The half-sine window in (5.4) was introduced in [3], where it was shown that it allows perfect dc reconstruction with only the low-pass subband. It was also shown in [3] that this property is a necessary condition for maximum asymptotic coding gain for an AR(1) signal with $\rho \to 1$. From here on, when we refer to the MLT we are implicitly assuming that the window $h(n)$ is defined by (5.4), with a minus sign.

In Fig. 5.1 we have a plot of the first four basis functions of the MLT, together with the magnitude frequency responses of all the basis functions. Since the subband filters are modulated versions of the same low-pass prototype, the shapes of the subband responses are the same. We note that the stopband attenuation is approximately 24 dB, which is slightly better than that of the LOTs that we discussed in Chapter 4

Unlike the LOT, the MLT basis functions have no symmetry, that is, their transfer functions are not linear phase. In applications such as speech coding and adaptive filtering, this is not a significant issue. However, when we are processing finite-length signals, such as in image coding, this lack of linear phase may lead to border effects, as we discuss in Section 5.4 and in Chapter 7.

Because of the modulation structure in the MLT basis functions in (5.1), there is a fast algorithm for the MLT computation, which is valid for any M. We will see in Section 5.4 the details of the fast algorithm. The coding performance of the MLT is also comparable to that of the LOT; we return to this issue in Section 5.5.

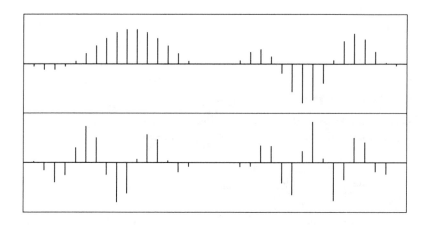

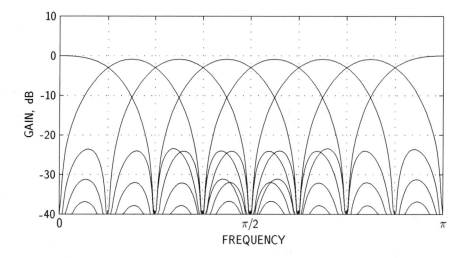

Figure 5.1. Top: the first four basis functions of the MLT, for $M = 8$; bottom: magnitude frequency responses of all basis functions.

5.2   Extended Lapped Transforms

The extended lapped transform (ELT) was developed as the answer to the following question: can the basis function of the MLT be made longer, with $L > 2M$, in such a way that perfect reconstruction is maintained? With longer basis functions, we would have better filtering performances, which may be useful in many applications. As the filters get closer to being ideal, the coding performance would also approach the optimum, for the given number of subbands [6].

The basis functions of the ELT are defined by the same cosine modulation function that was used for the MLT, but with a longer window:

$$p_{nk} = h(n)\sqrt{\frac{2}{M}} \cos\left[\left(n + \frac{M+1}{2}\right)\left(k + \frac{1}{2}\right)\frac{\pi}{M}\right] \tag{5.5}$$

for $k = 0, 1, \ldots, M-1$, and $n = 0, 1, \ldots, L-1$. We have already seen in the previous section that the basis functions above correspond to the general form of (3.22), with the modulation phases

$$\phi_k = \left(k + \frac{1}{2}\right)(N+1)\frac{\pi}{2} \tag{5.6}$$

where $L = NM$.

We saw in Section 3.3 that perfect reconstruction requires that the phases of the modulating cosines be related by

$$\phi_{k+1} - \phi_k = (2K+1)\frac{\pi}{2}$$

where K is an integer. Comparing the equation above with (5.6), we conclude that

$$N = 2K \quad \Rightarrow \quad L = 2KM \tag{5.7}$$

where K will be referred to as the *overlapping factor*. Thus, the length of the basis functions should be an even multiple of the number of subbands. Nevertheless, the above equation does not preclude the use of a window $h(n)$ such that $h(0) = h(1) = \cdots = h(r) = 0$, for some r. Thus, the actual length of the basis functions can be arbitrary. The main purpose of the restriction in (5.7) is to align the window to the correct phase with respect to the modulating cosines, in order to achieve perfect reconstruction. As it was the case with the MLT, the window is assumed to have the even symmetry of (5.2), that is,

$$h(2KM - 1 - n) = h(n)$$

Our problem now is to find a window $h(n)$, with length $L = 2KM$, that leads to perfect reconstruction, i.e., that turns the basis functions in (5.5) into a lapped transform. The conditions for perfect reconstruction were introduced in [7], and we present them in detail next.

5.2.1 Perfect Reconstruction

In order to obtain the PR property with the ELT, we should choose the window $h(n)$ such that the corresponding matrix $\mathbf{P}$ of the basis functions satisfies the generalized orthogonality conditions in (4.17). For this purpose, let us define the matrices $\mathbf{\Phi}_l$ and $\mathbf{H}_l$, of order M, by

$$[\mathbf{\Phi}_l]_{nk} \equiv \sqrt{\frac{2}{M}} \, \cos\left[\left(n + lM + \frac{M+1}{2}\right)\left(k + \frac{1}{2}\right)\frac{\pi}{M}\right] \tag{5.8}$$

and

$$\mathbf{H}_l \equiv \operatorname{diag}\left\{h(lM), \; h(lM + 1), \; \ldots, \; h(lM + M - 1)\right\}. \tag{5.9}$$

where $[\mathbf{\Phi}_l]_{nk}$ means the element at the nth row and kth column of the matrix $\mathbf{\Phi}_l$.

Grouping the square blocks $\mathbf{\Phi}_l$ and $\mathbf{H}_l$ into the $2KM \times M$ modulation matrix $\mathbf{\Phi}$ and the square window matrix $\mathbf{H}$ of order $2KM$, respectively, in the forms

$$\mathbf{\Phi}^T \equiv [\mathbf{\Phi}_0^T \; \mathbf{\Phi}_1^T \; \cdots \; \mathbf{\Phi}_{2K-1}^T]^T \tag{5.10}$$

and

$$\mathbf{H} \equiv \operatorname{diag}\{\mathbf{H}_0, \mathbf{H}_1, \ldots, \mathbf{H}_{2K-1}\} \tag{5.11}$$

we can write the ELT basis functions of (5.5) in matrix form as

$$\mathbf{P} = \mathbf{H}\,\mathbf{\Phi} \tag{5.12}$$

which is equivalent to

$$\mathbf{P}_i = \mathbf{H}_i \mathbf{\Phi}_i, \quad i = 0, 1, \ldots, 2K - 1 \tag{5.13}$$

where $\mathbf{P}_i$ is the ith square factor of $\mathbf{P}$, that is,

$$\mathbf{P}^T \equiv [\mathbf{P}_0^T \ \mathbf{P}_1^T \ \cdots \ \mathbf{P}_{2K-1}^T]$$

It is easy to verify that the submatrix $\mathbf{\Phi}_0$ of the modulation matrix $\mathbf{\Phi}$ can be written in the form

$$\mathbf{\Phi}_0 = \mathbf{C}\, \mathbf{D}_c - \mathbf{S}\, \mathbf{D}_s \tag{5.14}$$

where $\mathbf{C}$ and $\mathbf{S}$ are the orthogonal matrices that define the DCT-IV and DST-IV, respectively [see Section 1.2], and the matrices $\mathbf{D}_c$ and $\mathbf{D}_s$ are given by

$$\begin{aligned}
\mathbf{D}_c &\equiv \operatorname{diag}\left\{\cos\left[\frac{(2i+1)\pi}{4}\right]\right\} \\
&= \sqrt{1/2}\,\operatorname{diag}\{1 \ \text{-}1 \ \text{-}1 \ 1 \ \cdots\}
\end{aligned}$$

and

$$\begin{aligned}
\mathbf{D}_s &\equiv \operatorname{diag}\left\{\sin\left[\frac{(2i+1)\pi}{4}\right]\right\} \\
&= \sqrt{1/2}\,\operatorname{diag}\{1 \ 1 \ \text{-}1 \ \text{-}1 \ \cdots\}
\end{aligned}$$

Similarly, the submatrix $\mathbf{\Phi}_1$ of the modulation matrix $\mathbf{\Phi}$ can be written as

$$\mathbf{\Phi}_1 = -\mathbf{C}\, \mathbf{D}_c - \mathbf{S}\, \mathbf{D}_s \tag{5.15}$$

Using a basic relationship between the DCT-IV and DST-IV [8], which is

$$\operatorname{diag}\{1 \ \text{-}1 \ 1 \ \text{-}1 \ \cdots\}\, \mathbf{C} = \mathbf{S}\mathbf{J}$$

where $\mathbf{J}$ is the counter-identity matrix of (4.14), we arrive at

$$\mathbf{\Phi}_0 \mathbf{\Phi}_0^T = \mathbf{I} - \mathbf{J}$$

and

$$\mathbf{\Phi}_1 \mathbf{\Phi}_1^T = \mathbf{I} + \mathbf{J}$$

Furthermore, it holds that

$$\mathbf{\Phi}_0 \mathbf{\Phi}_1^T = \mathbf{\Phi}_1 \mathbf{\Phi}_0^T = \mathbf{0}$$

The other submatrices of $\mathbf{\Phi}$ are related to $\mathbf{\Phi}_0$ and $\mathbf{\Phi}_1$ by

$$\mathbf{\Phi}_{2i} = (-1)^i \, \mathbf{\Phi}_0 \tag{5.16}$$

and

$$\mathbf{\Phi}_{2i+1} = (-1)^i \, \mathbf{\Phi}_1 \tag{5.17}$$

Combining the properties above, we arrive at

$$\mathbf{\Phi}_i \mathbf{\Phi}_{i+2s}^T = (-1)^s [\mathbf{I} + (-1)^i \mathbf{J}]$$
$$\mathbf{\Phi}_i \mathbf{\Phi}_{i+2s+1}^T = \mathbf{0}$$

Then, the LT orthogonality conditions of (4.17) are

$$\sum_{i=0}^{N-1-l} \mathbf{P}_i \mathbf{P}_{i+l}^T = \sum_{i=0}^{2K-1-l} \mathbf{H}_i \, \mathbf{\Phi}_i \, \mathbf{\Phi}_{i+l}^T \, \mathbf{H}_{i+l}$$

$$= \begin{cases} \displaystyle\sum_{i=0}^{2K-l-1} \mathbf{H}_i \left[\mathbf{I} + (-1)^i \mathbf{J}\right] \mathbf{H}_{i+l}, & \text{if } l = 2s, \\ \mathbf{0}, & \text{otherwise} \end{cases} \tag{5.18}$$

We have seen in (4.17) that $\mathbf{P}$ is a lapped transform if and only if

$$\sum_{i=0}^{N-1-l} \mathbf{P}_i \mathbf{P}_{i+l}^T = \delta(l)\,\mathbf{I} \tag{5.19}$$

Substituting (5.19) into (5.18), an using the symmetry of the window, we obtain

$$\sum_{i=0}^{2K-2s-1} \mathbf{H}_i \, \mathbf{H}_{i+2s} = \delta(s)\,\mathbf{I} \tag{5.20}$$

for $s = 0, 1, \ldots, K - 1$. The scalar form of the above equation is

$$\sum_{i=0}^{2K-2s-1} h(n + iM) h(n + iM + 2sM) = \delta(s) \tag{5.21}$$

for $n = 0, 1, \ldots, M/2 - 1$, which was previously obtained in [7]. Thus, the set of nonlinear equations in (5.21) represents the necessary and sufficient conditions on the window $h(n)$ for the generation of an ELT.

For $K = 1$, the ELT becomes an MLT, and the PR conditions in (5.21) are identical to those in (5.3). For $K = 2$, the window conditions in (5.2) and (5.21) are

$$h(4M - 1 - n) = h(n), \ n = 0, 1, \ldots, 2M - 1,$$

$$h^2(n) + h^2(n + M) + h^2(n + 2M) + h^2(n + 3M) = 1, \ n = 0, 1, \ldots, M/2 - 1,$$

$$h(n)h(n + 2M) + h(n + M) + h(n + 3M) = 0, \ n = 0, 1, \ldots, M/2 - 1, \tag{5.22}$$

A smooth window that satisfies the above conditions is

$$h(n) = -\frac{1}{2\sqrt{2}} + \frac{1}{2} \cos\left[\left(n + \frac{1}{2}\right)\frac{\pi}{M}\right] \tag{5.23}$$

A parametrized family of windows can be generated, as suggested in [7], by

$$\begin{aligned}
h(n) &= c_{n0}\, c_{n1} \\
h(M - 1 - n) &= c_{n0}\, s_{n1} \\
h(M + n) &= s_{n0}\, c_{n1} \\
h(2M - 1 - n) &= -s_{n0}\, s_{n1}
\end{aligned} \tag{5.24}$$

for $n = 0, 1, \ldots, M/2 - 1$, where

$$\begin{aligned}
c_{n0} &\equiv \cos(\theta_{n0}) \\
c_{n1} &\equiv \cos(\theta_{n1}) \\
s_{n0} &\equiv \sin(\theta_{n0}) \\
s_{n1} &\equiv \sin(\theta_{n1})
\end{aligned} \tag{5.25}$$

with the angles being defined by

$$\theta_{n0} \equiv -\frac{\pi}{2} + \mu_{n+M/2}$$

$$\theta_{n1} \equiv -\frac{\pi}{2} + \mu_{M/2-1-n} \tag{5.26}$$

where

$$\mu_k \equiv \left[\left(\frac{1-\gamma}{2M}\right)(2k+1) + \gamma\right]\frac{(2k+1)\pi}{8M} \tag{5.27}$$

for $n = 0, 1, \ldots, M - 1$. The free parameter γ typically varies in the range $[0, 1]$, and the window $h(n)$ in (5.23) corresponds to the choice $\gamma = 1$.

In Figs. 5.2 to 5.4 we have the first four basis functions and the magnitude frequency responses of all basis functions of the ELT, for $M = 8$ and $K = 2$. They correspond, respectively, to the choices $\gamma = 1.0$, $\gamma = 0.7$, and $\gamma = 0.5$. We note that the parameter γ controls the trade-off between the stopband attenuation and transition width of the ELT frequency responses. For $K > 2$, we were not able to find simple parametric windows such as the one in (5.24)–(5.27). The design of the ELT window for $K > 2$ must rely on the optimization techniques described in Section 5.3.

5.2.2 Properties

There are three basic properties of the ELT basis functions that are a direct consequence of their modulated structure. First, the frequency responses of the synthesis filter bank are related to the window frequency response by

$$F_k(e^{j\omega}) = \frac{1}{\sqrt{2M}}\left[e^{-j\theta_k}H(e^{j(\omega-\omega_k)}) + e^{j\theta_k}H(e^{j(\omega+\omega_k)})\right]$$

where $\omega_k \equiv (k+1/2)\pi/M$ and $\theta_k \equiv \omega_k(M+1)/2$. Thus, except for $k = 0$ and $k = M - 1$, the magnitude responses of the analysis and synthesis filters are approximately given by

$$|F_k(e^{j\omega})| = |H_k(e^{j\omega})| \simeq \frac{1}{\sqrt{2M}}|H(e^{j(\omega-\omega_k)})| \tag{5.28}$$

where $|H(e^{j\omega})|$ is the magnitude frequency response of the window $h(n)$. Therefore, since we want each subband filter to have a bandwidth of approximately π/M, we see that the window should be a good low-pass filter with a cutoff frequency of about $\pi/2M$.

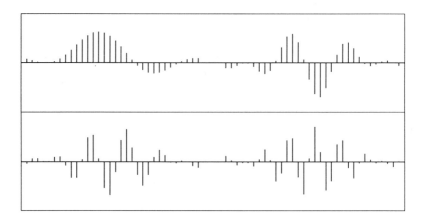

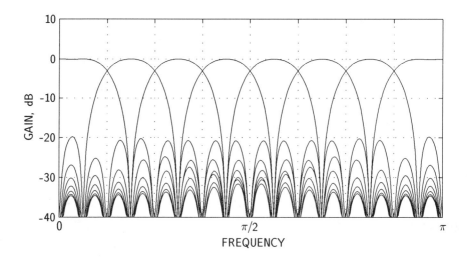

Figure 5.2. Top: the first four basis functions of the ELT, for $M = 8$, $K = 2$, and $\gamma = 1.0$; bottom: magnitude frequency responses of all basis functions.

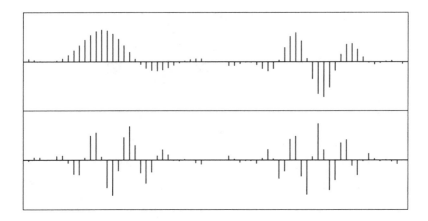

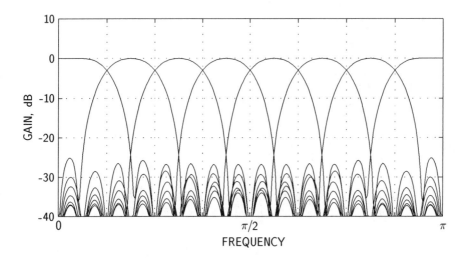

Figure 5.3. Top: the first four basis functions of the ELT, for $M = 8$, $K = 2$, and $\gamma = 0.7$; bottom: magnitude frequency responses of all basis functions.

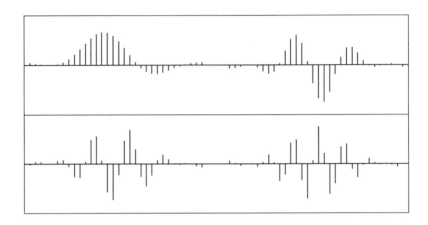

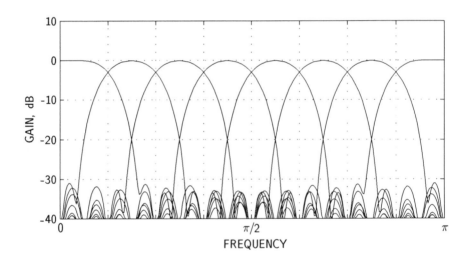

Figure 5.4. Top: the first four basis functions of the ELT, for $M = 8$, $K = 2$, and $\gamma = 0.5$; bottom: magnitude frequency responses of all basis functions.

The second interesting property of the ELT basis functions, which is easy to prove and is valid for even M, is

$$f_{M-1-k}(2KM - 1 - n) = (-1)^{k+K-M/2+1} (-1)^n f_k(n) \tag{5.29}$$

Thus, except for a plus or minus sign, half of the filter responses can be obtained from the other half by time reversal and multiplication by $(-1)^n$. Note that, for $M = 2$, We saw in Section 3.4 that this is precisely the property that defines the conjugate quadrature filters (CQF) [9]. Thus, the ELT can also be viewed as a generalization of the CQF concept for an arbitrary number of subbands.

The third property refers to the window $h(n)$. We saw in Section 3.4 that the each analysis or synthesis filter is a spectral factor of an Mth band filter, because of the perfect reconstruction property. With the ELT, PR means that the window $h(n)$ satisfies the conditions in (5.21). If we consider the mth polyphase component of the window, defined by

$$h_m(n) \equiv h(m + nM)$$

then the PR conditions can be written in the form

$$\sum_{i=0}^{2K-2s-1} h_m(i)\, h_m(i + 2s) = \delta(s) \tag{5.30}$$

for $m = 0, 1, \ldots, M/2 - 1$, and $s = 0, 1, \ldots, K - 1$. If we define a sequence $y_m(n)$ by

$$y_m(n) \equiv h_m(n) * h_m(-n)$$

then (5.30) becomes

$$y_m(2s) = \delta(s) \tag{5.31}$$

Thus, each polyphase component $h_m(n)$ of the window is a spectral factor of a half-band filter $y_m(n)$. Unfortunately, for different m we have different half-band filters. Therefore, this property is not useful for the design of ELT windows. In the next section we discuss a method for the design of $h(n)$.

5.3   Design Techniques

The MLT and ELT are characterized by the modulated basis functions of (5.5), which depend on the window $h(n)$, of length $L = 2KM$. When $K = 1$, the ELT becomes an MLT, and a good choice for $h(n)$ is the raised-sine window of (5.4). For $K = 2$, the parametrized window of (5.24)–(5.27) is also a good choice, and the free parameter γ can be used to control the trade-off between stopband attenuation and transition width of the frequency responses of the subband filters.

It is clear, then, that the PR conditions in (5.21) do not have a unique solution. For any M and K, we can always find an infinite number of solutions. To be more precise, we see that (5.21) imposes $KM/2$ equality conditions on the KM free coefficients $\{h(0), h(1), \ldots, h(KM - 1)\}$. Therefore, we have $KM/2$ degrees of freedom in choosing the window coefficients.

Given that many windows could be used, a natural question is: which is the best window? For coding applications, an optimal $h(n)$ would maximize the coding gain G_{TC}, a criterion that was used in Section 4.2 in one of the LOT design techniques. The disadvantage of such an approach is that the transform would be optimal only for the assumed signal covariance matrix. As we saw in Section 4.2, the LOT designed with basis on the QR decomposition, which was an attempt to obtain a good filtering performance, was more robust in terms of coding gain for varying signal spectra. There are also applications, such as transmultiplexing and adaptive filtering, where the coding gain is not a relevant measure, and the filtering performance is more important.

From the above discussion, we conclude that a good design criterion for the ELT would be to optimize the filtering performance, in such a way that the subband responses get as close as possible to the ideal bandpass responses. Furthermore, we know that no transform with M coefficients can have a coding gain higher than that of an M-band filter bank with ideal filters [6]. Therefore, subband filters designed for the best filtering performance should have coding gains close to the maximum for most signals.

Our design technique is based on the minimization of the stopband energy of the low-pass prototype, defined by

$$E_s \equiv \frac{1}{\pi} \int_{\omega_s}^{\pi} |H(e^{j\omega})|^2 \, d\omega \tag{5.32}$$

The frequency ω_s defines the beginning of the stopband, and it should be greater than $\pi/2M$, since the bandwidth of the low-pass prototype should be close to $\pi/2M$, as discussed above. For $\omega_s = \pi/M$, for example, a small value of E_s implies that there will be virtually no aliasing among subbands that are not adjacent to each other. This error measure is commonly used in filter design, including PR filter banks [10].

The design of $h(n)$ could be set as a constrained optimization problem. The set of independent variables would be $\{h(0), h(1), \ldots, h(KM-1)\}$, the objective function to be minimized would be the stopband energy E_s of (5.32), and there would be the $KM/2$ equality constraints of (5.21). Then, the augmented Lagrangian or another constrained optimization technique could be employed [11].

An alternative way of finding the best window is to formulate the optimization problem as unconstrained. In order to remove the constraints, we cannot use the window coefficients as the independent variables. A good set of unconstrained independent variables is the set of butterfly angles of the fast ELT structures that we will describe in the next section. This is because that structure is canonic, i.e., for any window that satisfies the PR conditions there is a corresponding set of angles, and any set of angles leads to a PR window.

Still, due to the highly nonlinear relationship that exists between the set of butterfly angles and the window coefficients, any optimization algorithm may get trapped in a local minimum. Therefore, it is of vital importance that a good initial set of angles be used. In order to solve this issue, we can formulate another optimization problem, in which we define an auxiliary window $w(n)$ as a sum of sinusoids, in the form

$$w(n) \equiv \sum_{i=0}^{K-1} a_i \cos \left[f_i \left(n - KM + \frac{1}{2} \right) \right] \quad n = 0, 1, \ldots, 2KM - 1 \qquad (5.33)$$

This window has the required even symmetry for any choice of the amplitudes a_i and frequencies f_i of the sinusoids. However, there is no guarantee that the window will satisfy the PR conditions. Thus, we perform first an unconstrained optimization in order to find the a_i and f_i that minimize the error in the PR conditions, defined by

$$E_{\text{PR}} \equiv \sum_{s=0}^{K-1} \sum_{n=0}^{M/2-1} \left[\sum_{i=0}^{2K-2s-1} h(n+iM) h(n+iM+2sM) - \delta(s) \right]^2 \qquad (5.34)$$

With K sinusoids, as suggested in (5.33), we can get $w(n)$ quite close to a PR window. Note that for $K = 1$ and $K = 2$ we can get exactly a PR window, according to (5.4) and (5.23). Then, the relationships in the next Section can be used to obtain an initial set of angles from $w(n)$. With these initial angles, the optimization of ξ is then performed. The steps of the algorithm are described below.

ELT Window Design Algorithm

Step 1: Set the variables a_i and f_i to random initial values, for $i = 0, 1, \ldots, K - 1$, and minimize E_{PR} as defined in (5.34), using an unconstrained minimization routine.

Step 2: Use (5.69), with $h(n) = w(n)$ to obtain the butterfly angles θ_{nr}, for $n = 0, 1, \ldots, M/2 - 1$, and $r = 0, 1, \ldots, K - 1$.

Step 3: Using the initial values of the angles θ_{nr} obtained in the previous step, minimize the stopband energy E_s as defined in (5.32), using an unconstrained optimization routine. In order to compute the window $h(n)$ at each step of the optimization algorithm, use (5.66).

Actually, Step 1 can be made to run faster if we use the results of a previous run of the algorithm with the same K, in order to initialize the parameters a_i and f_i. This is because a_i and f_i do not vary much with M.

A good choice for the unconstrained optimization routine that is required in Steps 1 and 3 is the Broyden-Fletcher-Goldfarb-Shanno (BFGS) algorithm, with scaling and periodic restarting [12]. The BFGS algorithm builds iteratively an estimate of the inverse Hessian matrix of the objective function; the inverse Hessian is used to correct the descent directions.

Because of the strong nonlinearities of the relationship between the window $h(n)$ and the butterfly angles, the optimization problem in Step 3 has a slow convergence, with an eigenvalue spread of the Hessian matrix that can be worse than five orders of magnitude. For $K = 4$ and $M > 16$, for example, it takes more than 200 iterations for convergence. The strong nonlinearities also mean that global optimality is not guaranteed. However, with the initial guess of butterfly angles that is obtained in Step 1, we can always expect to get good low-pass prototypes $h(n)$.

In Figs. 5.5 to 5.7 we have examples of ELTs designed with the algorithm described above, for $M = 8$ and $K = 1, 2, 3$, and 4. The stopband frequency ω_s was set to $0.8\pi/M$, π/M, and $1.2\pi/M$. The plots show only the frequency responses of the first three basis functions, so that we could better see the shapes of the stopband responses; no information is lost, since we know from (5.28) that the frequency responses are the virtually the same for all subbands.

As expected, we note from Figs. 5.5 to 5.7 that the stopband attenuation improves when we increase the overlapping factor K. When we vary ω_s, we control the trade-off between transition bandwidth and stopband attenuation. With $K = 4$ and $\omega_s = 0.8\pi/M$, for example, there is no overlap between the transition bands of nonadjacent subbands, for example. When $K = 1$, the ELT becomes equal to the MLT, except for the window. For the case $\omega_s = 1.2\pi/M$, the window designed with the algorithm described above becomes quite close to the raised-sine window of (5.4). For $K = 2$, the windows obtained with $\omega_s = 1.2\pi/M$ are approximately equal to the parametrized window of (5.24)–(5.27), for $\gamma = 0.3$. We have included tables of butterfly angles for the ELT in Appendix D, for different values of M, K, and ω_s. These angles can be used in the fast ELT structure that we present in the next section.

When choosing the best set of ELT basis functions in practice, other factors besides stopband attenuation or coding gain should be taken into account. For real-time speech processing, for example, the increase in the processing delay with K may be a limiting factor. We recall that the overall delay of the cascade of analysis and synthesis filter banks is $2KM - 1$.

For larger values of the overlapping factor K we obtain better filters, and so the coding gain also improves, as we will see in Section 5.5. In fact, the coding gains of the ELT get quite close to the upper bounds corresponding to ideal filters, which is an indication that the ELT window design algorithm has converged to the correct solutions.

5.4 Fast Algorithms

As we have seen up to now, the ELT has the desirable property of achieving perfect reconstruction, with a filtering performance that is as good as those of nonPR filter banks such as the pseudo-QMF bank. However, one of the most important characteristics of the ELT is that it can be efficiently computed, with a complexity close to that of block transforms.

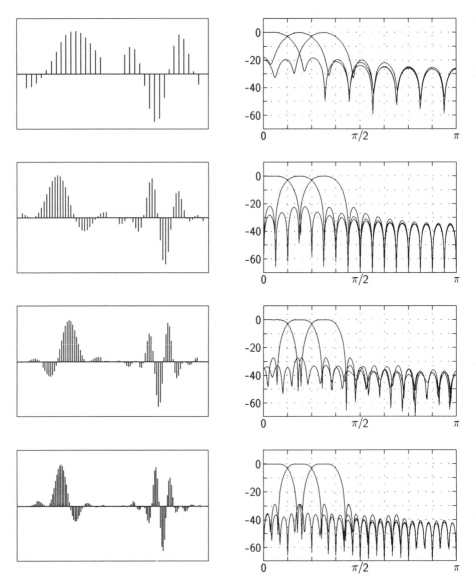

Figure 5.5. Examples of the ELT for $M = 8$ and $\omega_s = 0.8\pi/M$, for $K = 1$ (top), $K = 2$, $K = 3$, and $K = 4$ (bottom). Left: the first two basis functions; right: magnitude frequency responses of the first three basis functions.

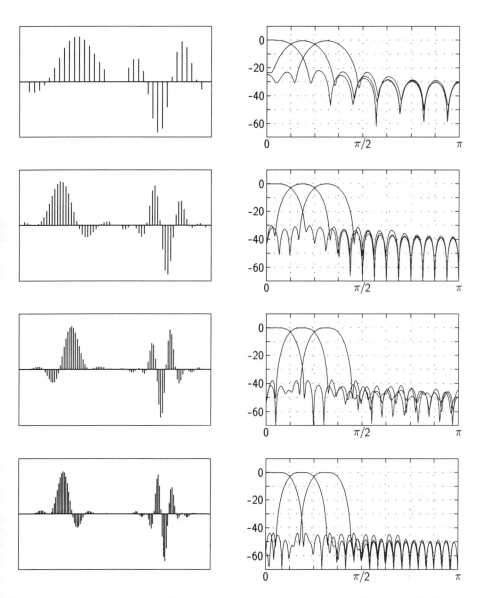

Figure 5.6. Examples of the ELT for $M = 8$ and $\omega_s = \pi/M$, for $K = 1$ (top), $K = 2$, $K = 3$, and $K = 4$ (bottom). Left: the first two basis functions; right: magnitude frequency responses of the first three basis functions.

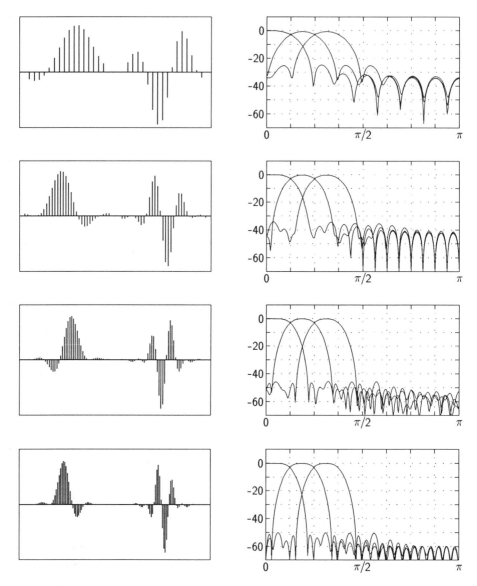

Figure 5.7. Examples of the ELT for $M = 8$ and $\omega_s = 1.2\pi/M$, for $K = 1$ (top), $K = 2$, $K = 3$, and $K = 4$ (bottom). Left: the first two basis functions; right: magnitude frequency responses of the first three basis functions.

When we look at the ELT definition in (5.5), and compare it with the DCT-IV definition of (1.23), it becomes clear that the modulating cosines of the ELT have the same frequencies of the basis functions of the DCT-IV. Therefore, we should expect the existence of a DCT-IV-based fast algorithm for the ELT. This is indeed the case, as we see in the following.

The value of the kth subband signal (or, equivalently, the kth transform coefficient) in the mth block is given by

$$
\begin{aligned}
X_k(m) &= h_k(n) * x(n) \,|_{n=mM} \\
&= \sum_{r=0}^{2KM-1} h_k(r) x(mM - r) \\
&= \sum_{r=0}^{2KM-1} h_k(2KM - 1 - r) x(mM - 2KM + 1 + r) \\
&= \sum_{r=0}^{2KM-1} f_k(r) x(mM - 2KM + 1 + r)
\end{aligned}
\tag{5.35}
$$

Defining the input block sequence by

$$
x_m(r) \equiv x(mM - 2KM + 1 + r)
\tag{5.36}
$$

and using the ELT definition in (5.5), we can put (5.35) in the form

$$
\begin{aligned}
X_k(m) &= \sum_{r=0}^{2KM-1} f_k(r) x_m(r) \\
&= \sqrt{\frac{2}{M}} \sum_{r=0}^{2KM-1} x_m(r) h(r) \cos\left[\left(n + \frac{M+1}{2}\right)\left(k + \frac{1}{2}\right)\frac{\pi}{M}\right]
\end{aligned}
\tag{5.37}
$$

In this equation, we have the ELT computation in a form quite similar to the block transforms in Chapter 2. The main difference is that the input block has $2KM$ samples, $x_m(0)$ to $x_m(2KM - 1)$, whereas the output block has only M samples, $X_0(m)$ to $X_{M-1}(m)$.

Using the change of variables $r = s + 3M/2$, and defining

$$
g(s) \equiv x_m(s + 3M/2) h(s + 3M/2)
\tag{5.38}
$$

we can write (5.37) in the form

$$
X_k(m) = -\sqrt{\frac{2}{M}} \sum_{s=-3M/2}^{(4K-3)M/2-1} g(s) \cos\left[\left(s + \frac{1}{2}\right)\left(k + \frac{1}{2}\right)\frac{\pi}{M}\right]
\tag{5.39}
$$

which is quite similar to the DCT-IV in (1.23), except that the input block is still $2KM$-sample long.

The key to the conversion of (5.39) into a block DCT-IV transform is to use the periodicity of the cosines to reduce the number of terms in the sum to M. Assuming that $g(s)$ is still defined as in (5.38), but with $g(s) = 0$ if $s < -3M/2$ or $s > (4K - 3)M/2$, we can write (5.39) in the form

$$X_k(m) = -\sqrt{\frac{2}{M}} \sum_{l=0}^{2K} \sum_{s=(l-2)M}^{(l-1)M-1} g(s) \cos\left[\left(s + \frac{1}{2}\right)\left(k + \frac{1}{2}\right)\frac{\pi}{M}\right] \tag{5.40}$$

In order to simplify the notation, let us define the DCT-IV basis functions by

$$\phi_k(s) \equiv \sqrt{\frac{2}{M}} \cos\left[\left(s + \frac{1}{2}\right)\left(k + \frac{1}{2}\right)\frac{\pi}{M}\right] \tag{5.41}$$

Also, splitting the first sum with $2K + 1$ terms in (5.40) into two sums for $l = 2s$ and $l = 2s + 1$, we get

$$X_k(m) = -\sum_{r=0}^{K} \sum_{s=(2r-2)M}^{(2r-1)M-1} g(s)\,\phi_k(s) - \sum_{r=0}^{K-1} \sum_{s=(2r-1)M}^{2rM-1} g(s)\,\phi_k(s) \tag{5.42}$$

Let us use two properties of the modulating cosines, which are simple to verify:

$$\phi_k(s + 2(r-1)M) = (-1)^{r-1}\phi_k(s) \tag{5.43}$$

and

$$\phi_k(2rM - 1 - s) = (-1)^r \phi_k(s) \tag{5.44}$$

Furthermore, let us apply the change of variables $v = s - (2r-2)M$ in the first term of (5.42), and the change $v = 2rM - 1 - s$ in the second term. Then, (5.42) becomes

$$X_k(m) = \sum_{v=0}^{M-1} u(v)\,\phi_k(v) \tag{5.45}$$

which we recognize as the length-M DCT-IV of the sequence

$$u(v) = \sum_{r=0}^{K}(-1)^r g(v + (2r-2)M) + \sum_{r=0}^{K-1}(-1)^{r-1}g(2rM - 1 - v) \tag{5.46}$$

Thus, we see that ELT coefficients of the mth block, $X_k(m)$, $k = 0, 1, \ldots, M-1$, can be obtained as the orthogonal DCT-IV of the M-sample sequence $u(v)$, $v = 0, 1, \ldots, M-1$. In order to obtain the sequence $u(v)$ from the $2KM$-sample input sequence $x_m(r)$, we must use (5.38) and (5.46).

A key point to the fast ELT implementation is the fact that the PR conditions in (5.21) imply in an orthogonal butterfly implementation of the mappings in (5.38) and (5.46). In order to see this fact, let us consider first the cases $K = 1$ and $K = 2$, and then generalize.

5.4.1 Fast ELT for $K = 1$ (MLT)

When $K = 1$, we have $L = 2M$, which corresponds to the MLT with a general window. The sequence $g(s)$ in (5.38) has length $2L$, with $g(s) = 0$ if $s \notin [-3M/2, M/2 - 1]$. Then, (5.46) can be simplified to

$$u(i) = \sum_{r=0}^{1} (-1)^r g(i + (2r-2)M) - g(-1-i)$$

from which it follows that

$$
\begin{aligned}
u(i + M/2) &= g(i - 3M/2) - g(-M/2 - 1 - i) \\
u(M/2 - 1 - i) &= -g(M/2 - 1 - i) - g(i - M/2)
\end{aligned}
$$

for $i = 0, 1, \ldots, M/2 - 1$. Using (5.38) and the window symmetry, we can rewrite the equation above as

$$
\begin{aligned}
u(i + M/2) &= x_m(i)\, h(i) - x_m(M - 1 - i)\, h(M - 1 - i) \\
u(M/2 - 1 - i) &= -x_m(2M - 1 - i)\, h(i) - x_m(i + M)\, h(M - 1 - i) \quad (5.47)
\end{aligned}
$$

Let us define the r-sample delay operator $\mathcal{D}_r\{\cdot\}$ by

$$\mathcal{D}_r\{y(n)\} \equiv y(n - r) \qquad (5.48)$$

Then, (5.47) can be written as

$$u(i + M/2) = \mathcal{D}_M\{x_m(i + M)\}\, h(i) - \mathcal{D}_M\{x_m(2M - 1 - i)\}\, h(M - 1 - i)$$

$$u(M/2 - 1 - i) = -x_m(2M - 1 - i)\, h(i) - x_m(i + M)\, h(M - 1 - i) \qquad (5.49)$$

We recall that the perfect reconstruction condition for $K = 1$ is equivalent to $h^2(i) + h^2(M - 1 - i) = 1$, in view of the window symmetry. Therefore, the mapping from $x_m(i)$ and $x_m(M - 1 - i)$ to $u(M/2 + i)$ and $u(M/2 - 1 - i)$ in (5.49) is an orthogonal butterfly with a z^{-M} delay. If we perform the butterfly after decimation of the input sequence, then the delay becomes z^{-1}.

The results in (5.45) and (5.49) map into the fast MLT flowgraph shown in Fig. 5.8. The corresponding inverse MLT flowgraph is also shown in Fig. 5.8. Again, we stress the property that the flowgraphs are orthogonal if and only if the window $h(n)$ satisfies the perfect reconstruction condition of (5.3). A similar structure for the fast MLT, with the same computational complexity, has been recently presented in [13].

We note that there are $M/2$ butterflies in Fig. 5.8. Each has the form

$$\begin{bmatrix} -\cos\theta & \sin\theta \\ \sin\theta & \cos\theta \end{bmatrix} \tag{5.50}$$

Because the matrix above is its own inverse, the butterflies in the inverse MLT are identical to those in the direct MLT.

If we call the butterfly angles in Fig. 5.8 θ_{00} in the outermost butterfly, and $\theta_{M/2-1,0}$ in the innermost butterfly, we see that the MLT window is related to the butterfly angles by

$$\begin{aligned} h(n) = h(2M - 1 - n) &= -\cos\theta_{n0} \\ h(n + M) = h(M - 1 - n) &= -\sin\theta_{n0} \end{aligned} \tag{5.51}$$

which guarantee the PR property. The raised-sine window of (5.4) corresponds to the following choice of angles:

$$\theta_{n0} = \frac{\pi}{2} - \frac{\pi}{2M}\left(n + \frac{1}{2}\right) \tag{5.52}$$

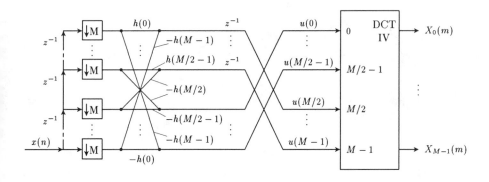

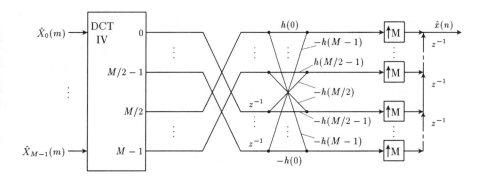

Figure 5.8. Flowgraph of the fast ELT for $K = 1$, which corresponds to the MLT. Top: direct transform, or analysis filter bank; bottom: inverse transform, or synthesis filter bank.

5.4.2   Fast ELT for $K = 2$

When $K = 2$, that is, $L = 4M$, the sequence $g(s)$ in (5.38) has length $4L$, with $g(s) = 0$ if $s \notin [-3M/2, 5M/2 - 1]$. The sequence $u(n)$ in (5.46) can be written as

$$u(i) = \sum_{r=0}^{2}(-1)^r g(i + (2r - 2)M) + \sum_{r=0}^{1}(-1)^{r-1}g(2rM - 1 - i)$$

As we did in the case $K = 1$, let us consider the two halves of $u(n)$ simultaneously, with indices $n = i + M/2$ and $n = M/2 - 1 - i$. We have

$$
\begin{aligned}
u(i + M/2) &= g(i - 3M/2) - g(i + M/2) \\
&\quad -g(-M/2 - 1 - i) + g(3M/2 - 1 - i) \\
u(M/2 - 1 - i) &= -g(M/2 - 1 - i) + g(5M/2 - 1 - i) \\
&\quad -g(i - M/2) + g(3M/2 + i)
\end{aligned}
$$

for $i = 0, 1, \ldots, M/2 - 1$. Again using the window symmetry and (5.38), we obtain

$$
\begin{aligned}
u(i + M/2) &= x_m(i)\,h(i) \\
&\quad -x_m(i + 2M)\,h(2M - 1 - i) \\
&\quad -x_m(M - 1 - i)\,h(M - 1 - i) \\
&\quad +x_m(3M - 1 - i)\,h(i + M) \\
u(M/2 - 1 - i) &= -x_m(2M - 1 - i)\,h(2M - 1 - i) \\
&\quad +x_m(4M - 1 - i)\,h(i) \\
&\quad -x_m(i + M)\,h(i + M) \\
&\quad +x_m(i + 3M)\,h(M - 1 - i) \tag{5.53}
\end{aligned}
$$

We wish to express the two samples $u(i + M/2)$ and $u(M/2 - 1 - i)$ as functions of only two samples of $x_m(\cdot)$, preferably $x_m(i + 3M)$ and $x_m(4M - 1 - i)$, since they are the most recent in the block. Using the delay operator $\mathcal{D}_r\{\cdot\}$ of (5.48), we can rewrite (5.53) in the form

$$
\begin{aligned}
u(i + M/2) &= \mathcal{D}_{3M}\{x_m(i)\}\,h(i) \\
&\quad -\mathcal{D}_M\{x_m(i + 3M)\}\,h(2M - 1 - i) \\
&\quad -\mathcal{D}_{3M}\{x_m(4M - 1 - i)\}\,h(M - 1 - i) \\
&\quad +\mathcal{D}_M\{x_m(4M - 1 - i)\,h(i + M)
\end{aligned}
$$

$$
\begin{aligned}
u(M/2 - 1 - i) = \ & -\mathcal{D}_{2M}\{x_m(4M - 1 - i)\}\,h(2M - 1 - i) \\
& +x_m(4M - 1 - i)\,h(i) \\
& -\mathcal{D}_{2M}\{x_m(i + 3M)\}\,h(i + M) \\
& +x_m(i + 3M)\,h(M - 1 - i)
\end{aligned}
\tag{5.54}
$$

Now let us define the window $h(n)$ as in (5.24), that is,

$$
\begin{aligned}
h(i) &= c_{i0}\,c_{i1} \\
h(M - 1 - i) &= c_{i0}\,s_{i1} \\
h(i + M) &= s_{i0}\,c_{i1} \\
h(2M - 1 - i) &= -s_{i0}\,s_{i1}
\end{aligned}
$$

for $i = 0, 1, \ldots, M/2 - 1$, where the c_{ir} and s_{ir} were defined in (5.25).

We have seen in Section 5.2 that such a choice guarantees perfect reconstruction. Furthermore, let us define an auxiliary M-point sequence $y(i)$ by

$$
\begin{aligned}
y(i) &\equiv \mathcal{D}_{2M}\{-c_{i1}\,x(i + 3M) + s_{i1}\,x(4M - 1 - i)\} \\
y(M - 1 - i) &\equiv s_{i1}\,x(i + 3M) + c_{i1}\,x(4M - 1 - i)
\end{aligned}
\tag{5.55}
$$

Then, it is easy to show that (5.54) is equivalent to

$$
\begin{aligned}
u(i + M/2) &\equiv \mathcal{D}_M\{-c_{i0}\,y(i) + s_{i0}\,y(M - 1 - i)\} \\
u(M/2 - 1 - i) &\equiv s_{i0}\,y(i) + c_{i0}\,y(M - 1 - i)
\end{aligned}
\tag{5.56}
$$

From (5.55) and (5.56), we see that the sequence $u(\cdot)$ can be obtained from the sequence $x_m(\cdot)$ by means of a cascade of two butterfly stages, intermixed with delays. The corresponding flowgraphs for the direct and inverse ELTs are shown in Fig. 5.9. Each butterfly is labeled only with its angle, which defines the butterfly transmittances according to (5.50).

We note that the fast ELT structure for $K = 2$ is equal to the fast ELT structure for $K = 1$, plus an extra set of butterflies also in the form in (5.50), and a set of z^{-2} delays. A natural question that arises now is: can we add more butterfly stages and z^{-2} delays to the flowgraphs in Fig. 5.9 in order to obtain fast ELTs for $K > 2$? Fortunately, the answer is yes, as we see next.

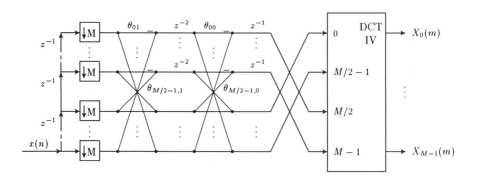

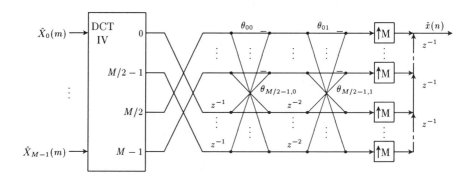

Figure 5.9. Flowgraph of the fast ELT for $K = 2$. Top: direct transform, or analysis filter bank; bottom: inverse transform, or synthesis filter bank.

5.4.3   General Case

Looking at the ELT flowgraphs for $K = 1$ and $K = 2$ in Figs. 5.8 and 5.9, we see that the general ELT flowgraph should have the forms indicated in Fig. 5.10. The matrices $\mathbf{D}_0$ to $\mathbf{D}_{K-1}$ are the butterfly stages. According to (5.50), they have the form

$$\mathbf{D}_k \equiv \left[\begin{array}{cc} -\mathbf{C}_k & \mathbf{S}_k\mathbf{J} \\ \mathbf{J}\mathbf{S}_k & \mathbf{J}\mathbf{C}_k\mathbf{J} \end{array} \right] \tag{5.57}$$

where

$$\begin{aligned} \mathbf{C}_k &\equiv \operatorname{diag}\{\cos\theta_{0k}, \cos\theta_{1k}, \ldots, \cos\theta_{M/2-1,k}\} \\ \mathbf{S}_k &\equiv \operatorname{diag}\{\sin\theta_{0k}, \sin\theta_{1k}, \ldots, \sin\theta_{M/2-1,k}\}. \end{aligned} \tag{5.58}$$

In order to verify that the generalized structures of Fig. 5.10 generate all possible ELTs for a given overlapping factor K, let us compare the inverse ELT flowgraph of Fig. 5.10 with the lossless polyphase component matrix interpretation of Fig. 3.27. Assuming that the structure does correspond to an ELT for some K, let us add another stage of z^{-2} delays and a butterfly matrix $\mathbf{D}_K$. Then, the polyphase component matrix $\mathbf{E}_{K+1}$ of the new structure will be related to the polyphase component matrix $\mathbf{E}_K$ of the original structure by

$$z^{-[2(K+1)-1]}\,\mathbf{J}\tilde{\mathbf{E}}_{K+1}(z) = \mathbf{D}_K \left[\begin{array}{cc} \mathbf{I} & \mathbf{0} \\ \mathbf{0} & z^{-2}\mathbf{I} \end{array} \right] z^{-(2K-1)}\,\mathbf{J}\tilde{\mathbf{E}}_K(z)$$

Substituting (5.57) in the above equation, we obtain

$$\mathbf{J}\tilde{\mathbf{E}}_{K+1}(z) = \left[\begin{array}{cc} -\mathbf{C}_k & \mathbf{S}_k\mathbf{J} \\ \mathbf{J}\mathbf{S}_k & \mathbf{J}\mathbf{C}_k\mathbf{J} \end{array} \right] \left[\begin{array}{cc} z^2\mathbf{I} & \mathbf{0} \\ \mathbf{0} & \mathbf{I} \end{array} \right] \mathbf{J}\tilde{\mathbf{E}}_K(z) \tag{5.59}$$

Now let us write the polyphase component matrix as a function of the window coefficients. In order to do so, we recall from (4.13) that

$$\tilde{\mathbf{E}}(z) = \mathbf{E}^T(z^{-1}) = \sum_{l=0}^{2K-1} z^{(2K-1-l)}\,\mathbf{J}\,\mathbf{P}_l \tag{5.60}$$

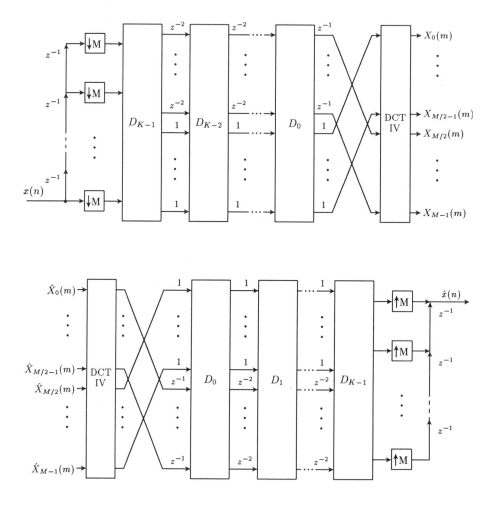

Figure 5.10. Flowgraph of the fast ELT for any K. Top: direct transform, or analysis filter bank; bottom: inverse transform, or synthesis filter bank.

where $\mathbf{P}_l$ is defined in (5.13). Using the basic properties of the modulation submatrices in (5.16) and (5.17), we can write (5.60) as

$$\mathbf{J}\tilde{\mathbf{E}}_K(z) = \sum_{r=0}^{K-1} z^{2(K-r)-1} (-1)^r [\mathbf{H}_{2r}^K \mathbf{\Phi}_0 + z^{-1}\mathbf{H}_{2r+1}^K \mathbf{\Phi}_1] \qquad (5.61)$$

where H_i^K denotes the ith block of the window matrix [see 5.11], for an ELT structure with K butterfly stages.

The submatrices $\mathbf{\Phi}_0$ and $\mathbf{\Phi}_1$ have rank $M/2$, and can be written in the forms:

$$\mathbf{\Phi}_0 = \begin{bmatrix} \mathbf{I} \\ -\mathbf{J} \end{bmatrix} \mathbf{\Lambda}_0 \qquad (5.62)$$

and

$$\mathbf{\Phi}_1 = \begin{bmatrix} \mathbf{I} \\ \mathbf{J} \end{bmatrix} \mathbf{\Lambda}_1 \qquad (5.63)$$

where $\mathbf{\Lambda}_0$ and $\mathbf{\Lambda}_1$ can be easily computed from (5.14), and are $M/2 \times M$ matrices.

Furthermore, we can break the diagonal window submatrices in four blocks, in the form

$$\mathbf{H}_i^K = \begin{bmatrix} \mathbf{U}_i^K & \mathbf{0} \\ \mathbf{0} & (-1)^i \mathbf{J}\mathbf{V}_i^K\mathbf{J} \end{bmatrix} \qquad (5.64)$$

where $\mathbf{U}_i^K$ and $\mathbf{V}_i^K$ are also diagonal matrices. Again, the superscript K denotes not exponentiation, but the overlapping factor for which the window was computed. Note that we need to work only with $\mathbf{H}_0^K$ to $\mathbf{H}_{K-1}^K$, because the window symmetry implies

$$\mathbf{H}_{2K-1-i}^K = \mathbf{J}\mathbf{H}_i^K\mathbf{J}$$

Using (5.62)–(5.64), we can write (5.61) as

$$\mathbf{J}\tilde{\mathbf{E}}_K(z) = \sum_{r=0}^{K-1} z^{2(K-r)-1}(-1)^r \left(\begin{bmatrix} \mathbf{U}_{2r}^K \\ -\mathbf{J}\mathbf{V}_{2r}^K \end{bmatrix} \mathbf{\Lambda}_0 + z^{-1} \left(\begin{bmatrix} \mathbf{U}_{2r+1}^K \\ -\mathbf{J}\mathbf{V}_{2r+1}^K \end{bmatrix} \mathbf{\Lambda}_1 \right) \right) \qquad (5.65)$$

When we substitute (5.59) into (5.65), we get an expression for $\tilde{\mathbf{E}}_{K+1}(z)$ that is precisely in the form of (5.65), and leads to the following recursion for the window components:

$$
\begin{aligned}
\mathbf{U}_i^{K+1} &= -\mathbf{C}_K \mathbf{U}_i^K + \mathbf{S}_K \mathbf{V}_{i-2}^K \\
\mathbf{V}_i^{K+1} &= -\mathbf{S}_K \mathbf{U}_i^K - \mathbf{C}_K \mathbf{V}_{i-2}^K
\end{aligned}
\tag{5.66}
$$

for $i = 0, 1, \ldots, K$. The recursion works as follows: starting with the initial conditions $\mathbf{U}_0^0$ and $\mathbf{V}_{-2}^0$ that we will define next, we compute $\mathbf{U}_0^1$ and $\mathbf{V}_0^1$, which completely define the window for $K = 1$. Next, we compute $\mathbf{U}_0^2$, $\mathbf{U}_1^2$, $\mathbf{V}_0^2$, and $\mathbf{V}_0^2$, which define the window for $K = 2$. We continue until we reach the desired value of K. When $i = K$ in (5.66), we use $\mathbf{U}_K^K \equiv 0$ for $K \geq 2$.

According to the relationships between the window coefficients and the butterfly angles that we obtained before for $K = 1$ and $K = 2$, it is clear that the initial conditions for the recursion in (5.66) are

$$
\mathbf{U}_0^0 = \mathbf{I}, \quad \mathbf{V}_{-2}^0 = \mathbf{0}
\tag{5.67}
$$

which are equivalent to

$$
\mathbf{U}_1^1 = -\mathbf{C}_0^1, \quad \mathbf{V}_0^1 = -\mathbf{S}_0^1
\tag{5.68}
$$

These are precisely the same relationships that we obtained in (5.51) for $K = 1$.

The relationships in (5.66) can be easily inverted, leading to

$$
\begin{aligned}
\mathbf{U}_i^K &= -\mathbf{C}_K \mathbf{U}_i^{K+1} - \mathbf{S}_K \mathbf{V}_i^{K+1} \\
\mathbf{V}_i^K &= \mathbf{S}_K \mathbf{U}_{i+2}^{K+1} - \mathbf{C}_K \mathbf{V}_{i+2}^{K+1}
\end{aligned}
\tag{5.69}
$$

for $i = 0, 1, \ldots, K - 1$. This recursion is important because it allows us to compute the butterfly angles from a given window. This is necessary, for example, in Step 2 of the ELT window design algorithm of the previous section.

In the recent literature, the ELT has also been referred to as the perfect-reconstruction cosine modulated filter bank. Other fast implementations for the ELT, similar to the one described in this section, were recently presented in [14, 15].

5.4.4   Computational Complexity

The fast ELT structures of Fig. 5.10 are based on a DCT-IV block transforms plus K butterfly stages. Assuming that the DCT-IV is computed using the algorithm described in Section 2.5, which requires a total of $M + (M/2)\log_2 M$ multiplications and $(3M/2)\log_2 M$ additions, and that each butterfly is computed with three multiplications and three additions [16] [see also Fig. 2.14], the total complexity of the fast ELT would be $(M/2)(3K + \log_2 M + 2)$ multiplications and $(M/2)(3K + 3\log_2 M)$ additions.

The complexity can be reduced further by a slight modification of the butterfly matrices $\mathbf{D}_k$. We see in Fig. 5.10 that the cascade of the matrices $\mathbf{D}_k$ leads to a cascade connection of butterflies and delays. Therefore, we could scale all the coefficients in butterflies $\mathbf{D}_1$ to $\mathbf{D}_{K-1}$ in such a way that all the diagonal entries would be equal to 1 or -1. In order to compensate for these modifications, the inverse scaling would be applied to $\mathbf{D}_0$. Thus, instead of using the butterfly matrices $\mathbf{D}_k$, we could use the matrices $\overline{\mathbf{D}}_k$, defined by

$$\overline{\mathbf{D}}_k \equiv \text{diag}\{\mathbf{C}_k^{-1}, \mathbf{JC}_k^{-1}\mathbf{J}\}\,\mathbf{D}_k, \ k = 1, \ldots, K-1$$

$$\overline{\mathbf{D}}_0 \equiv \text{diag}\{\prod_{k=1}^{K-1} \mathbf{C}_k, \mathbf{J} \prod_{k=1}^{K-1} \mathbf{C}_k\mathbf{J}\}\,\mathbf{D}_0 \tag{5.70}$$

Then, $\overline{\mathbf{D}}_0$ would still require three multiplications and three additions per butterfly, but each $\overline{\mathbf{D}}_k$ would need only two multiplications and two additions per butterfly. The final computational complexity of the fast ELT would be

$$\mu(M) = \frac{M}{2}(2K + \log_2 M + 3)$$

$$\alpha(M) = \frac{M}{2}(2K + 3\log_2 M + 1) \tag{5.71}$$

For $M = 2$, the scaling factors in (5.70) can be absorbed in the single butterfly that implements the DCT-IV, and thus we would save one extra multiplication and one extra addition. In Table 5.1 we have a summary of the fast ELT complexity.

We note that the computational complexity of the ELT is close to that of block transforms, and thus much lower than those of other filter banks such as the QMF bank [17] or the PR structures based on canonical factorizations [10]. For example, for $M = 32$ and $K = 2$, the number of operations (multiplications and additions) per input sample is 16, for the ELT. As a comparison, the DCT requires 9.1 operations per sample, and the PR structure of [10] needs 104 operations per sample.

M	K=1		K=2		K=3		K=4		K=5	
	MUL	ADD	MUL	ADD	MUL	ADD	MUL	ADD	MUL	ADD
2	5	5	7	7	9	9	11	11	13	13
4	14	18	18	22	22	26	26	30	30	34
8	32	48	40	56	48	64	56	72	64	80
16	72	120	88	136	104	152	120	168	136	184
32	160	288	192	320	224	352	256	384	288	416
64	352	672	416	736	480	800	544	864	608	928
128	768	1536	896	1664	1024	1792	1152	1920	1280	2048
256	1664	3456	1920	3712	2176	3968	2432	4224	2688	4480
512	3584	7680	4096	8192	4608	8704	5120	9216	5632	9728
1024	7680	16896	8704	17920	9728	18944	10752	19968	11776	20992

Table 5.1. Computational complexity of the fast ELT algorithm.

5.4.5 ELTs of Finite-Length Signals

In Section 4.3 we saw that the fast LOT flowgraph had to be slightly modified for the computation of the LOT of finite-length signals, such as rows or columns of an image. We had to modify the LOT flowgraph for the first and last blocks of the signal in order to preclude the basis functions to extend beyond the signal boundaries.

Because the LOT basis functions have linear phase, we saw that the an orthogonal LOT flowgraph for finite length signals was obtained simply by assuming that the signal was evenly reflected at its boundaries. With the MLT and ELT, however, such a reflection does not make sense because the basis functions are not symmetric. If we assume symmetric signal reflection, the corresponding signal flowgraphs for the MLT and ELT would contain nonorthogonal factors that would lead to artifacts in the reconstructed image in coding applications.

In order to keep the orthogonality of the total flowgraph for finite length signals, we have tried to introduce orthogonal matrices such that the basis functions corresponding to the blocks at the borders of the signal were as smooth as possible, and as close as possible to being polyphase normalized, i.e., a purely DC signal should lead to a strong energy only in the low-pass subband signals. We did that only for

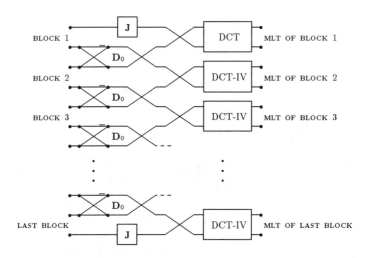

Figure 5.11. Flowgraph of the fast MLT for a finite length signal. Note that a DCT replaces the DCT-IV in the first block. The direct transform is computed from left to right, and the inverse transform from right to left.

$K = 1$ and $K = 2$ because in image processing applications there is no advantage in using larger values of K, as we see in Chapter 7.

The flowgraphs of the corrected ELT for finite-length signals are shown in Figs. 5.11 and 5.12, for $K = 1$ (MLT) and $K = 2$, respectively. In Fig. 5.12 we have assumed that the input signal contains five blocks, but it is easy to generalize it to any number of blocks. In that structure, note that only the ELT of the middle block is computed according to a flowgraph that corresponds exactly to the fast ELT structures of Fig. 5.9. The ELTs of the first two and the last two blocks are modified.

It is interesting to note that in both cases one of the DCT-IV blocks is replaced by a DCT. This leads to a more compact representation of dc signals, i.e., the structures become closer to being polyphase normalized. We note that the LOT flowgraph of Section 4.3 was exactly polyphase-normalized, so that even at the boundary blocks a dc signal would have only the first LOT coefficient being nonzero. With the ELT structures corrected for finite-length signals in Figs. 5.11 and 5.12, polyphase normalization is not exact. Thus, for a dc input signal, the first and last

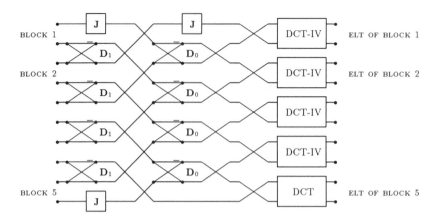

Figure 5.12. Flowgraph of the fast ELT for a finite length signal, for $K = 2$. Note that a DCT replaces the DCT-IV in the last block. The direct transform is computed from left to right, and the inverse transform from right to left.

K blocks will have all of their ELT coefficients different from zero, but the low-pass coefficient will be the largest. In Appendix C we have a set of "C" programs for the computation of the MLT and the ELT. Those programs are implementations of all the fast ELT flowgraphs that we have discussed in this section.

5.5   Coding Performance

As we did in Chapter 4, it is interesting to compare the transform coding gains G_{TC} [see Section 2.3] of the ELT with those of the DCT and LOT. Although G_{TC} is a theoretical measure, the differences in G_{TC} from one transform to another serve to predict reasonably well the expected improvement in SNR, for a given bit rate.

In Tables 5.2 to 5.4 we have the coding gains of the ELT, DCT, LOT, and the upper bound corresponding to ideal filter banks. We have used the same set of values for the overlapping factor K and the stopband frequency ω_s as we did in Figs. 5.5 to 5.7. In all tables, the coding gains for the LOT correspond to the type-II LOT, which is the only one that is fast-computable for any number of subbands.

TRANSFORM	NUMBER OF SUBBANDS, M					
	2	4	8	16	32	64
DCT	5.06	7.58	8.83	9.47	9.79	9.95
LOT	–	7.94	9.22	9.77	9.99	10.07
ELT K=1, $\omega_s = 0.8\pi/M$	3.82	6.46	8.25	9.28	9.76	9.95
ELT K=2, $\omega_s = 0.8\pi/M$	5.68	8.32	9.43	9.88	10.04	10.09
ELT K=3, $\omega_s = 0.8\pi/M$	5.82	8.41	9.51	9.92	10.06	10.09
ELT K=4, $\omega_s = 0.8\pi/M$	5.89	8.52	9.57	9.96	10.08	10.11
ELT K=1, $\omega_s = \pi/M$	4.89	7.49	8.92	9.63	9.93	10.04
ELT K=2, $\omega_s = \pi/M$	5.76	8.40	9.50	9.92	10.06	10.10
ELT K=3, $\omega_s = \pi/M$	5.88	8.50	9.57	9.95	10.08	10.11
ELT K=4, $\omega_s = \pi/M$	5.88	8.52	9.58	9.96	10.08	10.11
ELT K=1, $\omega_s = 1.2\pi/M$	5.48	8.01	9.22	9.78	10.00	10.08
ELT K=2, $\omega_s = 1.2\pi/M$	5.75	8.40	9.51	9.92	10.06	10.11
ELT K=3, $\omega_s = 1.2\pi/M$	5.86	8.48	9.56	9.95	10.08	10.11
ELT K=4, $\omega_s = 1.2\pi/M$	5.87	8.50	9.57	9.95	10.08	10.11
IDEAL	5.96	8.59	9.62	9.97	10.08	10.11

Table 5.2. Coding gain in dB of several transforms, for an AR(1) signal with $\rho = 0.95$.

In all cases, we note a significant reduction in the coding gain when we reduce ω_s from π/M to $0.8\pi/M$; this leads us to the conclusion that we should avoid using ω_s lower than π/M. For $K = 1$ and $\omega_s = 1.2\pi/M$, the ELT becomes an MLT, and the window is approximately equal to the raised-sine window of (5.4). We see that the MLT coding gains are slightly higher than those of the LOT. Since the MLT has a lower computational complexity than the type-II LOT, we conclude that the MLT should be preferred over the LOT for applications requiring a large number of subbands, such as speech coding and adaptive filtering. We return to this LOT versus MLT comparison in Chapter 7.

Considering the ELTs for $\omega_s = 1.2\pi/M$, we see that the coding gain generally jumps when we move from the DCT to the ELT with $K = 1$, and then the gain increases slowly with K. For $K = 4$, the gain becomes quite close to the maximum that could be achieved with an ideal filter bank. When we recall from Section 5.3 that ELTs with $\omega_s = 1.2\pi/M$ and $K = 4$ can achieve stopband attenuations in the order of 50 dB, we conclude that in the great majority of applications it would not be necessary to work with overlapping factors greater than four.

Transform	Number of Subbands, M					
	2	4	8	16	32	64
DCT	3.49	5.35	7.49	9.25	10.49	11.38
LOT	–	5.96	8.67	10.17	11.30	12.13
ELT K=1, $\omega_s = 0.8\pi/M$	3.13	6.49	8.13	9.35	10.44	11.50
ELT K=2, $\omega_s = 0.8\pi/M$	3.69	7.55	9.68	11.02	11.79	12.38
ELT K=3, $\omega_s = 0.8\pi/M$	3.82	8.12	10.21	11.27	12.03	12.44
ELT K=4, $\omega_s = 0.8\pi/M$	3.82	8.25	10.47	11.55	12.15	12.62
ELT K=1, $\omega_s = \pi/M$	3.68	6.85	8.75	10.11	11.09	11.95
ELT K=2, $\omega_s = \pi/M$	3.89	7.78	9.96	11.26	11.97	12.49
ELT K=3, $\omega_s = \pi/M$	3.87	8.16	10.44	11.54	12.17	12.62
ELT K=4, $\omega_s = \pi/M$	3.86	8.23	10.55	11.62	12.20	12.64
ELT K=1, $\omega_s = 1.2\pi/M$	3.89	6.85	9.10	10.64	11.52	12.21
ELT K=2, $\omega_s = 1.2\pi/M$	3.95	7.69	10.00	11.34	12.03	12.52
ELT K=3, $\omega_s = 1.2\pi/M$	3.89	8.04	10.37	11.54	12.16	12.63
ELT K=4, $\omega_s = 1.2\pi/M$	3.88	8.10	10.44	11.56	12.18	12.66
IDEAL	3.57	8.50	11.62	11.99	12.30	12.76

Table 5.3. Coding gain in dB of several transforms, for a male speech signal.

Transform	Number of Subbands, M					
	2	4	8	16	32	64
DCT	1.78	3.09	4.27	5.27	6.11	6.63
LOT	–	3.84	5.07	5.98	6.56	7.54
ELT K=1, $\omega_s = 0.8\pi/M$	2.47	3.88	4.92	5.31	6.26	7.21
ELT K=2, $\omega_s = 0.8\pi/M$	2.93	4.40	5.67	6.14	6.97	7.79
ELT K=3, $\omega_s = 0.8\pi/M$	3.27	4.60	5.80	6.24	7.13	7.90
ELT K=4, $\omega_s = 0.8\pi/M$	3.29	4.63	5.95	6.36	7.20	8.00
ELT K=1, $\omega_s = \pi/M$	2.50	4.12	5.26	5.71	6.56	7.43
ELT K=2, $\omega_s = \pi/M$	2.89	4.46	5.80	6.27	7.05	7.83
ELT K=3, $\omega_s = \pi/M$	3.18	4.60	5.96	6.37	7.16	7.92
ELT K=4, $\omega_s = \pi/M$	3.20	4.62	6.01	6.40	7.20	7.96
ELT K=1, $\omega_s = 1.2\pi/M$	2.41	4.18	5.43	5.97	6.72	7.52
ELT K=2, $\omega_s = 1.2\pi/M$	2.77	4.44	5.82	6.31	7.05	7.80
ELT K=3, $\omega_s = 1.2\pi/M$	3.10	4.56	5.95	6.38	7.14	7.90
ELT K=4, $\omega_s = 1.2\pi/M$	3.13	4.58	5.98	6.39	7.17	7.94
IDEAL	3.30	4.63	6.05	6.43	7.33	8.16

Table 5.4. Coding gain in dB of several transforms, for a female speech signal.

When we compare the performances of the DCT for $M = 64$ and the ELT with $M = 16$ and $K = 2$, we see that the coding gains are similar. In both cases, the length of the analysis and synthesis filters would be equal to 64, and thus the processing delay would also be the same. However, the total computational complexity of the analysis and synthesis filter banks would be 1,412 operations for the DCT and 448 operations for the ELT. Also, adaptive quantization of 16 ELT coefficients would require much less computational effort than the quantization of 64 DCT coefficients. Furthermore, we should not forget that the smooth decay of the ELT basis functions at their boundaries means an absence of the blocking effects. In the speech coding experiments reported in Chapter 7 we will verify in practice the superior ELT performance.

As a final point, it is interesting to note that the ELT coding gain is insensitive to the sign of the correlation coefficient of an AR(1) source. This property does not hold for the DCT and LOT. For example, for $M = 8$ and $\rho = 0.95$, the DCT has a coding gain of 8.83 dB, and the LOT and ELT with $K = 1$ have their coding gains equal to 9.22 dB. For $\rho = -0.95$, the coding gain of the DCT drops drastically to 5.09 dB, the LOT gain drops to 8.92 dB, and the ELT gain stays at 9.22 dB. Although most signal sources have positive correlation coefficients, there are practical cases where the correlation coefficient may be negative, for example for signals resulting from differencing operators or signal portions containing higher amounts of high-frequency components [18].

5.6   Summary

In this chapter we have discussed in detail the theory and properties of the modulated lapped transform (MLT) and extended lapped transform (ELT), which are LTs whose basis functions are generated from sinusoidal modulations of a low-pass prototype. In Section 5.1 we studied the MLT, which is a particular case of the ELT, with basis functions of the same length as those of the LOT, i.e., twice the number of subbands.

The more general family of ELTs was the subject of Section 5.2, where we derived the perfect reconstruction conditions on the window $h(n)$. The design of a good window that satisfies the PR conditions was discussed in Section 5.3, where a two nonlinear optimizations problems were involved. The goal of the optimizations were two find a window that led to the minimum stopband energy, with the stopband being defined as $[\omega_s, \pi]$. From the results in Sections 5.3 and 5.4, it seems that $\omega_s = 1.2\pi/M$ seems to be an adequate choice for most applications. The optimization of the window is not necessary for $K = 1$ and $K = 2$. For $K = 1$, which corresponds to the MLT, the raised-sine window of (5.4) can be used. For $K = 2$, the parametrized window of (5.24) to (5.27) can be used with γ between 0.3 and 0.5.

By choosing appropriately the overlapping factor of the ELT, we can make it more suitable to the application at hand. In image coding, for example, where block transforms are traditional used, ELTs with $K = 1$ or $K = 2$ should be adequate, and both will virtually eliminate the blocking effects and will also lead to better coding gains, as we saw in Section 5.5. For adaptive filtering, transmultiplexing, and other applications where larger stopband attenuations may be necessary, ELTs with larger overlapping factors K may be used.

The ELT would not be a good substitute for block transforms or for other filter banks if it were not for its fast computational algorithms, which we described in Section 5.4. For example, for $M = 16$, the total number of arithmetic operations required by the ELT with $K = 2$ is approximately twice the number of operations required by the DCT. For applications such as adaptive filtering of telephone-line modem signals or real-time speech coding, for example, the computation of the ELT would take less than one tenth of the computing power of an inexpensive digital signal processor.

For coding applications, the theoretical coding gain of the ELT is an important performance measure. As we saw in Section 5.5, the ELT can lead to significantly higher coding gains than block transforms such as the DCT. For a wide class of signals, an ELT with $K = 4$ has a coding gain that is close enough to the maximum that could be achieved with an ideal filter bank. Therefore, in applications such as transform coding of speech, we could replace a DCT by an ELT with a smaller number of subbands and get the same coding performance, and no blocking effects.

References

[1] Princen, J. P., and A. B. Bradley, "Analysis/synthesis filter bank design based on time domain aliasing cancellation," *IEEE Trans. Acoust., Speech, Signal Processing*, vol. ASSP-34, Oct. 1986, pp. 1153–1161.

[2] Princen, J. P., A. W. Johnson, and A. B. Bradley, "Sub-band/transform coding using filter bank designs based on time domain aliasing cancellation," *IEEE Intl. Conf. Acoust., Speech, Signal Processing*, Dallas, pp. 2161–2164, Apr. 1987.

[3] Malvar, H. S., "Lapped transforms for efficient transform/subband coding," *IEEE Trans. Acoust., Speech, Signal Processing*, vol. 38, June 1990, pp. 969–978.

[4] Kovacevic, J., D. J. Le Gall, and M. Vetterli, "Image coding with windowed modulated filter banks," *IEEE Intl. Conf. Acoust., Speech, Signal Processing*, Glasgow, Scotland, pp. 1949–1952, May 1989.

[5] Isabelle, S. H., and J. S. Lim, "On modulated filter banks for image coding applications," *IEEE Digital Signal Processing Workshop*, New Paltz, NY, Sep. 1990.

[6] Pearlman, W. A., "Performance bounds for subband coding," in *Subband Image Coding*, (Woods, J. W., ed.), ch. 1, Boston, MA: Kluwer, 1991.

[7] Malvar, H. S., "Modulated QMF filter banks with perfect reconstruction," *Electron. Lett.*, vol. 26, no. 13, June 1990, pp. 906–907.

[8] Wang, Z., "Fast algorithms for the discrete W transform and for the discrete Fourier transform," *IEEE Trans. Acoust., Speech, Signal Processing*, vol. ASSP-32, Aug. 1984, pp. 803–816.

[9] Smith, M. J. T., and T. P. Barnwell III, "Exact reconstruction techniques for tree-structured subband coders," *IEEE Trans. Acoust., Speech, Signal Processing*, vol. ASSP-34, June 1986, pp. 434–441.

[10] Vaidyanathan, P. P., T. Q. Nguyen, T. Saramäki, and Z. Doğanata, "Improved technique for design of perfect reconstruction FIR QMF banks with lossless polyphase matrices," *IEEE Trans. Acoust., Speech, Signal Processing*, vol. 37, July 1989, pp. 1042–1056.

[11] Bertsekas, D., *Constrained Optimization and Lagrange Multiplier Methods*, New York: Academic Press, 1982.

[12] Luenberger, D. G., *Linear and Non-Linear Programming*, Reading, MA: Addison-Wesley, 2nd ed., 1984. Chap. 9.

[13] Duhamel, P., Y. Mahieux, and J. P. Petit, "A fast algorithm for the implementation of filter banks based on time domain aliasing cancelation," *IEEE Intl. Conf. Acoust., Speech, Signal Processing*, Toronto, Canada, pp. 2209–2212, May 1991.

[14] Ramstad, T. A., and J. P. Tanem, "Cosine-modulated analysis-synthesis filterbank with critical sampling and perfect reconstruction," *IEEE Intl. Conf. Acoust., Speech, Signal Processing*, Toronto, Canada, pp. 1789–1792, May 1991.

[15] Koilpillai, R. D., and P. P. Vaidyanathan, "New results on cosine-modulated filter banks satisfying perfect reconstruction," *IEEE Intl. Conf. Acoust., Speech, Signal Processing*, Toronto, Canada, pp. 1793–1796, May 1991.

[16] Wang, Z., "On computing the discrete Fourier and cosine transforms," *IEEE Trans. Acoust., Speech, Signal Processing*, vol. ASSP-33, Oct. 1985, pp. 1341–1344.

[17] Crochiere, R. E., and L. R. Rabiner, *Multirate Digital Signal Processing*, Englewood Cliffs, NJ: Prentice-Hall, 1983.

[18] Akansu, A. N., and R. A. Haddad, "On asymmetrical performance of discrete cosine transform," *IEEE Trans. Acoust., Speech, Signal Processing*, vol. 38, Jan. 1990, pp. 154–156.

Chapter 6

Hierarchical Lapped Transforms

Multiresolution signal decomposition is the main topic of this chapter. The filter banks that we have discussed so far, including lapped transforms and block transforms, were uniform. Thus, all basis functions had the same length, and their frequency responses had the same bandwidth. With many practical signals, however, high-frequency phenomena such as edges and pulses tend to have short durations, whereas low-frequency components have long durations. Hence, it is often useful to perform a subband decomposition such that the low-frequency subbands have narrower widths (and thus longer impulse responses) than the high-frequency subbands. In this way, the lower subbands have fine frequency resolution and coarse time resolution, whereas the higher subbands have coarse frequency resolution and fine time resolution. With this multiresolution approach, we are able to analyze signal features occurring at many different time and frequency scales.

The basic tool for multiresolution signal decomposition is the nonuniform filter bank, which we will discuss in Section 6.1. Hierarchical transforms, which are equivalent to nonuniform filter banks from a transform viewpoint, will be studied in Section 6.2. When we use a lapped transform as the building block for a hierarchical transform, we have the advantage of low computational complexity. By gaining in computation, we do not lose in filtering performance. In fact, hierarchical lapped transforms are efficient wavelet decompositions, as we will see in Section 6.3. Keeping with the philosophy of previous chapters, we will compare the performances of several nonuniform filter banks, in terms of transform coding gain, in Section 6.4.

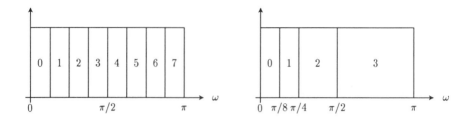

Figure 6.1. Example of uniform and nonuniform subband divisions. Left: uniform division into eight bands; right: four nonuniform subbands in an octave scale. In both cases, the low-pass subband has the same width of $\pi/8$.

6.1   Nonuniform Filter Banks

In the uniform filter banks that we studied in Chapter 3, the signal spectrum was divided into equal-width subbands, as shown in Fig. 6.1. When we want to perform a multiresolution signal analysis, we need a nonuniform subband division. This is particularly true when the signals of interest have wide bandwidths, in terms of the ratio of the largest to the lowest frequency components. In Fig. 6.1 we have also an example of a nonuniform subband division into four bands. The subbands are divided in an octave scale: band 0 covers one-eighth of the signal spectrum, bands 0 and 1 cover one-fourth, and so on.

6.1.1   Direct Form

In the subband analysis and synthesis system of Fig. 3.7, we assumed that the subband filters $H_0(z)$ to $H_{M-1}(z)$ and $F_0(z)$ to $F_{M-1}(z)$ had bandwidths equal to π/M, approximately. Thus, in order to maintain the total sampling rate, each subband was decimated by a factor of M. If we allow each analysis and synthesis filter to have a different bandwidth, we need also different sampling rates for each subband, as shown in Fig. 6.2. The only restriction that is implicitly imposed in Fig. 6.2 is that the analysis and synthesis filters for the same subband have the same bandwidth, since the decimation and interpolation factors for each subband are the same.

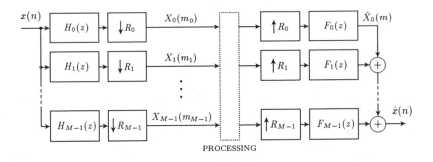

Figure 6.2. Nonuniform filter bank with several sampling rates.

Because of the different sampling rates of each subband in Fig. 6.2, we may need up to M different time indices for the subband signals, namely m_0 to m_{M-1}. With the uniform filter banks that we have studied so far, we had only one index m for the subband signals, and m could be referred to as a block counter. In the nonuniform filter bank of Fig. 6.2 we can still keep this block processing viewpoint, but to each subband there is a different sampling rate, and hence a different block size. According to Fig. 6.2, the kth subband is processed in blocks of length R_k, since R_k is the subsampling factor for that band.

Let us assume that the bandwidth of the kth subband in Fig. 6.2 is W_k radians. A reasonable assumption on the filters is that their passbands cover completely the input spectrum, that is,

$$\sum_{k=0}^{M-1} W_k = \pi \tag{6.1}$$

Generally, we want to preserve the total sampling rate. Thus, the average number of samples per unit time (or space) in all subbands should be equal to the input sampling rate. This is automatically achieved if each subband is critically decimated: if we choose the subsampling rates according to

$$R_k = \frac{\pi}{W_k} \tag{6.2}$$

then it follows from (6.1) that

$$\sum_{k=0}^{M-1} \frac{1}{R_k} = 1 \qquad\qquad (6.3)$$

which guarantees that the total sampling rate is preserved. In most cases we can assume that the subsampling factors R_k are integers, and so the subband widths are submultiples of π.

6.1.2   Tree Structures

A nonuniform filter bank could be implemented according to the structure of Fig. 6.2. However, the problems of aliasing cancellation and perfect reconstruction would be quite difficult to address, if we analyze that structure as a set of parallel filters. In fact, there is no simple direct design method for PR filter banks in the parallel structure of Fig. 6.2. Depending on the desired subband widths, though, a tree of PR filters can be used to build a nonuniform PR filterbank, as we will see next.

In many practical cases we are interested in nonuniform filter banks in which the subband widths are chosen in a regular way. An example is the octave scale subdivision of Fig. 6.1. For that choice of subbands, the general structure of Fig. 6.2 assumes the form shown in Fig. 6.3. Note that there are only three different time indices for the subbands signals in Fig. 6.3, because subbands number 0 and 1 have the same subsampling factor. A tree structure that leads to the same nonuniform subband splitting is shown in Fig. 6.4. It is based on the simple idea that we can generate the octave scale subdivision in a recursive form. We start by dividing the input spectrum in two bands, with the first stage of Fig. 6.4. Then, we split the low-pass band into two subbands, with the second stage of Fig. 6.4. We keep splitting the resulting low-pass subband with more stages, until the final low-pass subband has the desired width.

The tree structure of Fig. 6.4 has several advantages. First, because each stage splits its input in two subbands, we need only two filters, $H_{\mathrm{LP}}(z)$ and $H_{\mathrm{HP}}(z)$, with low-pass and high-pass responses, respectively. Second, the same filters can be used in all stages, and this leads to simplified hardware or software implementations. Third, ensuring aliasing cancellation and perfect reconstruction is straightforward: we simply pick QMF or CQF filters [see Chapter 3] for $H_{\mathrm{LP}}(z)$ and $H_{\mathrm{HP}}(z)$.

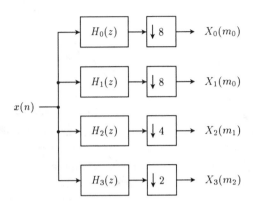

Figure 6.3. Nonuniform analysis filter bank with four octave-spaced subbands.

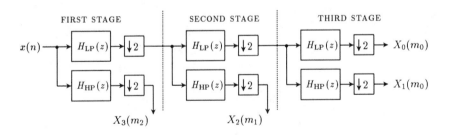

Figure 6.4. Tree structure for the implementation of four octave-spaced subbands. $H_{LP}(z)$ and $H_{HP}(z)$ are low-pass and high-pass filters, respectively, with cutoff frequencies at $\pi/2$.

A disadvantage of the tree structure is that we lose the generality of individually choosing each subband filter. For example, $H_0(z)$ must be the cascade connection of several $H_{LP}(z)$ filters and interpolators. If the length of $h_{LP}(n)$ is L, the length of $h_0(n)$ is $L_0 = (2^S - 1)(L - 1) + 1$, where S is the number of stages in the tree. Thus, of the L_0 coefficients of $h_0(n)$, there are only L degrees of freedom in choosing them.

In Chapter 3 we mentioned that the use of tree structures for the implementation of uniform filter banks was not recommended, mainly for a large number of subbands, because of the aliasing build-up that comes from subsequent subsampling [see Fig. 3.20]. This problem is much less severe in the nonuniform tree structure of Fig. 6.4, because only the low-pass branches are subject to further subsampling. Any path that goes through a high-pass filter ends immediately after that filter and its decimator, as shown in Fig. 6.4. As we will see in the Section 6.3, spurious sidelobes as those in Fig. 3.20 do not occur in the filter bank of Fig. 6.4.

A compromise between flexibility of nonuniform subband divisions and aliasing build-up can be achieved if we allow some of the high-pass filter outputs to go through further stages of subband splitting. The key point is that the depth of a path in the tree should be kept short, if it passes through a high-pass filter. This idea is incorporated in the generalized hierarchical lapped transform, which is discussed later in the next section.

6.2   Hierarchical Transforms

A hierarchical transform (HT) is a nonuniform filter bank, as in Fig. 6.2, in which the synthesis filters are seen as basis functions of a transform operator, and the analysis filters are equal to the time-reversed synthesis filters. Throughout this chapter, we reserve the term HT only for those nonuniform filter banks in which the corresponding basis functions satisfy an appropriate set of orthogonality conditions. If the filter length for a particular subband is equal to its decimation factor R_k, that subband operates as a block transform, with block length R_k. If the filter length for the kth subband is greater than R_k, that subband operates as a lapped transform.

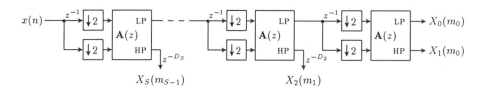

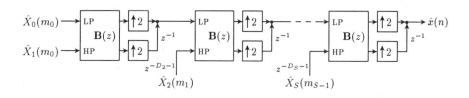

Figure 6.5. Hierarchical lapped transform (HLT). Top: direct transform (analysis filter bank); bottom: inverse transform (synthesis filter bank). $\mathbf{A}(z)$ is lossless, and $\mathbf{B}(z) = z^{N-1}\tilde{\mathbf{A}}(z)$. The delays D_2 to D_S are necessary to equalize the processing delay for all subbands.

6.2.1 HLT with Octave Band Splitting

When each of the stages in Fig. 6.4 is a lapped transform, we call the resulting nonuniform filter bank a *hierarchical lapped transform*. In order to visualize the structure as a cascade of transform operators, we can use the polyphase component matrix implementation of Fig. 3.27. Considering that S stages are used in the tree, the HLT assumes the form in Fig. 6.5. The polyphase component matrix of the analysis filters is $\mathbf{A}(z)$, with the low-pass and high-pass outputs denoted in Fig. 6.5 by LP and HP, respectively. For the synthesis filters, the polyphase component matrix is $\mathbf{B}(z)$.

According to the results in Section 3.4, we know that perfect reconstruction is possible in the structure of Fig. 6.5, with identical analysis and synthesis filters (within time reversal), i.e., with

$$\mathbf{B}(z) = z^{-(N-1)}\tilde{\mathbf{A}}(z) = z^{-(N-1)}\mathbf{A}^T(z^{-1}) \tag{6.4}$$

where $N - 1$ is the order of the polyphase component matrices. We are assuming that both $\mathbf{A}(z)$ and $\mathbf{B}(z)$ are causal, that is, each entry has no positive powers of z. We recall that the corresponding filters $h_{\mathrm{LP}}(n)$ and $h_{\mathrm{HP}}(n)$ have their lengths equal to $L = 2N$. The delays D_2 to D_S are necessary to equalize the overall processing delay among the subbands, as we will discuss next.

The necessary and sufficient condition for perfect reconstruction is that $\mathbf{A}(z)$ must be lossless:

$$\mathbf{A}(z)\,\tilde{\mathbf{A}}(z) = \mathbf{I} \tag{6.5}$$

The PR property of the HLT in Fig. 6.5 is easy to verify, when the polyphase component matrices satisfy (6.4) and (6.5). Assume that $\tilde{X}_k(m) = X_k(m), \forall\ k$ in Fig. 6.5, i.e., that all outputs of the direct transform are connected directly to the inputs of the inverse transform. Then, the last stage of the direct HLT and the first stage of the inverse HLT could be removed and replaced by a delay of $2N - 1$ samples because of the PR property of the two-band filter bank of each stage. This delay is precisely what must be introduced in subband number 2, and so $D_2 = N$.

Following the same reasoning, the next-to-last stage of the direct HLT and the second stage of the inverse HLT can be removed, and replaced by their equivalent overall delay, which is $6N - 3$ samples. Thus, the total delay to subband number 3 must be also $6N - 3$, and so $D_3 = 3N - 1$. It is clear, then, that the subband delays should be given by

$$D_k = (2^{k-1} - 1)N - 2^{k-2} + 1 \tag{6.6}$$

for $k = 2, 3, \ldots, S$. At the end of this process, we will have removed all stages, and $\hat{x}(n)$ will be a replica of $x(n)$, with a total delay of

$$D_{\mathrm{TOT}} = (2^S - 1)(2N - 1) = L_0 - 1 \tag{6.7}$$

where, as before, L_0 is the length of the equivalent filter for the low-pass subband.

If the polyphase component matrix $\mathbf{A}(z)$ is memoryless, i.e., if $\mathbf{A}(z) = \mathbf{A}$, then each stage in the HLT is an orthogonal block transform of length two. For example, if $\mathbf{A}$ is a DCT matrix, then the HLT would be a cascade of DCT operators. Cascades of DCT operators have been applied to progressive image transmission, but in a nonorthogonal fashion [1–3]. If $N > 1$, that is, if $\mathbf{A}(z)$ has order greater than zero,

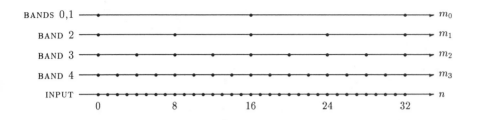

Figure 6.6. The several sampling rates of the HLT, for $S = 4$.

then each stage is a lapped transform. Since each stage is a two-band filter bank, the filters are also conjugate quadrature filters (CQF) [see Section 3.4].

With S stages in the hierarchical transform structure of Fig. 6.5, the number of subbands is $M = S + 1$, and the widths of each subband are

$$W_k = \begin{cases} 2^{-S}\pi, & \text{if } k = 0 \\ 2^{-(S-k+1)}\pi, & \text{if } k = 1, 2, \dots, S \end{cases} \tag{6.8}$$

which correspond to an octave scale, i.e., the bandwidths increase by factors of two. The effective subsampling rate for each band is

$$R_k = \begin{cases} 2^S, & \text{if } k = 0 \\ 2^{S-k+1}, & \text{if } k = 1, 2, \dots, S \end{cases} \tag{6.9}$$

According to (6.8) and (6.9), we see that the HLT provides S different time-frequency resolutions. Thus, each subband signal provides information on the input signal at a different scale, except for $X_0(m_0)$ and $X_1(m_0)$, which have the same time and frequency scales. The several sampling rates of the HLT are illustrated in Fig. 6.6, for a tree with four stages.

6.2.2 Generalized HLT

The HLT of Fig. 6.5 generates a grid of sampling rates according to Fig. 6.6, and a corresponding octave scale subband splitting. As we will see in Section 6.4 and also in Chapter 7, for signal coding applications there may be other nonuniform subband divisions that will lead to a better system performance. A more general

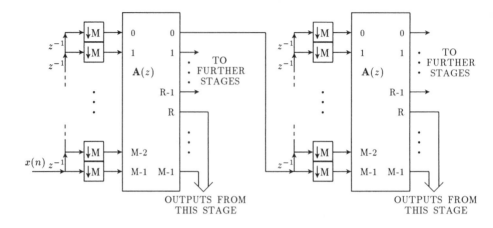

Figure 6.7. Generalized HLT, analysis filter bank (direct transform).

HLT can be obtained if we use more than two subbands in each stage, as suggested in [4] and shown in Fig. 6.7.

With the generalized HLT structure, there are M subbands in each stage, and R of them are brought up in the hierarchy for further subband splitting. By varying M, R, and S, we can generate many different nonuniform subband splitting patterns. For example, with $M = 4$, $R = 2$, and $S = 2$ we have the subband division in Fig. 6.8.

Another generalization of the original HLT structure of Fig. 6.5 would be to allow some of the high-pass subbands to go through further levels of subband splitting. In this way, we could actually obtain the same subband division of Fig. 6.8, for example. However, as we have discussed in Section 3.3, further subsampling of high-pass subbands may lead to undesirable sidelobes in the frequency responses, as shown in Fig. 3.20. For this reason, we believe that we should always try to use an HLT with the largest possible number of subbands per stage; this will also minimize the number of stages in the tree.

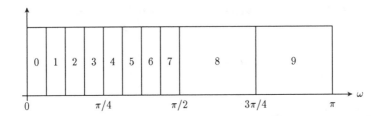

Figure 6.8. Example of nonuniform subband division in an HLT with $M = 4$, $R = 2$, and $S = 2$.

For the generalized HLT, we use M to denote the number of subbands per stage. The total number of subbands, which we will denote by M_T, is given by

$$M_T = (M - R) + (M - R)R + \cdots + (M - R)R^{S-2} + MR^{S-1}$$

or

$$M_T = \frac{(M - R)(R^{S-1} - 1)}{R - 1} + MR^{S-1} \tag{6.10}$$

where S is the number of stages, or the depth of the tree. In fact, we could carry the generalization one step further, and allow each transform to have a different number of subbands. Furthermore, the number of subbands R that are brought up to the next level could also vary from stage to stage. In this way, the possibilities of creating complex nonuniform subband divisions are virtually limitless. As long as we use a lapped transform in each stage, we are guaranteed to achieve perfect reconstruction, no matter how we build the HLT tree.

6.3 Connections with Wavelet Transforms

The HLT has the purpose of performing a multiresolution signal decomposition by means of a perfect reconstruction nonuniform filter bank. In fact, the HLT can be viewed as a discrete wavelet transform [5, 6] with a faster implementation algorithm than a tree of lattice structures (which is usually employed to implement orthogonal wavelet transforms). In this section we study the connection between HLTs and wavelet transforms in detail.

6.3.1   The HLT as a Multiresolution Transform

When we have a uniform filter bank, i.e., when $R_k = M, \forall k$, then the filter bank of
Fig. 6.2 is identical to that in Fig. 3.7. We recall from Section 3.2 that the subband
signals are given by the analysis equation

$$X_k(m) = \sum_{n=-\infty}^{\infty} x(n) h_k(mM - n) \tag{6.11}$$

and the reconstructed signal is given, assuming no modification of the subband
signals, by the synthesis equation

$$\hat{x}(n) = \sum_{k=0}^{M-1} \sum_{m=-\infty}^{\infty} X_k(m) f_k(n - mM) \tag{6.12}$$

From the results in Chapters 3 to 5, we know that we can always choose FIR analysis
and synthesis filters such that $\hat{x}(n) = x(n - D)$, where D is a processing delay. This
is what we have been referring to as perfect reconstruction (PR). When we use the
lapped transforms of Chapters 4 and 5 for the analysis and synthesis filters in (6.11)
and (6.12), we have fast algorithms to move from the original signal to the subband
signals, and viceversa.

With PR FIR analysis and synthesis filters, (6.11) and (6.12) provide a useful
time-frequency signal decomposition. Instead of looking at the original signal $x(n)$,
which is a function of the time index n, we look at the subband signals $X_k(m)$.
The time index m now refers to blocks of M samples, whereas k is a frequency
index. For example, if at a given block index m the k−th subband signal has a
large magnitude, we know that the signal $x(n)$ has a strong sinusoidal component
near the frequency $k\pi/M$, and near the time instant $n = mM$. Thus, we can locate
sinusoidal features both in time and in frequency.

The main disadvantage of the uniform filter bank, as we have discussed in Section 6.1 is that the time-frequency representation has a fixed resolution in both time
and frequency. We cannot resolve in time features with durations shorter than the
block length M, and this would drive us to choose small values of M. Furthermore,
we cannot resolve frequency components that are closer than π/M, and this would
lead us into choosing large values of M. In seismic signal processing, for example [7],
it may be impossible to find a value of M that will satisfy the time and frequency
resolution requirements.

With the nonuniform filter bank of Fig. 6.2, we can move from single resolution to multiresolution. For each subband, we simply choose the bandwidth W_k and the block length R_k that leads us to the best tradeoff between time and frequency resolutions for that band. We can choose different widths for each subband, as long as (6.1) to (6.3) are satisfied. The analysis and synthesis equations for the nonuniform filter bank of Fig. 6.2 are

$$X_0(m_0) = \sum_{n=-\infty}^{\infty} x(n)\, h_0(m_0 R_0 - n)$$

$$X_k(m_{k-1}) = \sum_{n=-\infty}^{\infty} x(n)\, h_k(m_{k-1} R_k - n) \tag{6.13}$$

and

$$\hat{x}(n) = \sum_{m_0=-\infty}^{\infty} X_0(m_0)\, f_0(n - m_0 R_0)$$

$$+ \sum_{k=1}^{M-1} \sum_{m_k=-\infty}^{\infty} X_k(m_{k-1})\, f_k(n - m_{k-1} R_k) \tag{6.14}$$

In the equations above we are assuming that the first two subbands are indexed by m_0, whereas the other subbands are indexed by m_1, m_2, etc. This is because we want to use the HLT structure of Fig. 6.5 to implement the nonuniform filter bank.

The analysis and synthesis equations can be viewed as direct and inverse transform operators, if the synthesis filters are the basis functions and the analysis filters are the time-reversed basis functions. In fact, if we use the HLT structure of Fig. 6.5, we know that the analysis and synthesis filters are all generated by a pair of two-band PR filters, which we will refer to as $f_{LP}(n)$ and $f_{HP}(n)$. The polyphase components of $f_{LP}(n)$ and $f_{HP}(n)$ are the entries of $\mathbf{B}(z)$ in Fig. 6.5.

6.3.2   Equivalent Subband Filters

We have seen in Section 6.2 that the HLT of Fig. 6.5 corresponds to a nonuniform filter bank, with subsampling factors (or block sizes) given by (6.9). In order to find the equivalent subband filter for the kth subband in Fig. 6.5, we can use the noble identities of Chapter 3. The results can be written in a simple recursive form, which we will present next. Let us define the functions $\phi_1(n)$ and $\psi_1(n)$ by

$$\phi_1(n) \equiv f_{LP}(n)$$
$$\psi_1(n) \equiv f_{HP}(n) \tag{6.15}$$

Furthermore, let us also define the $1\!:\!2$ interpolation operator $\mathcal{I}_2\{\cdot\}$

$$\mathcal{I}_2\{x(n)\} = \begin{cases} x(n/2), & n \text{ even} \\ 0, & n \text{ odd} \end{cases}$$

Let us consider the recursion formulas

$$\begin{aligned} \phi_{i+1}(n) &= \mathcal{I}_2\{\phi_i(n)\} * \phi_1(n) \\ \psi_{i+1}(n) &= \mathcal{I}_2\{\psi_i(n)\} * \phi_1(n) \end{aligned} \tag{6.16}$$

where $*$ is the convolution operator. Then, it is clear from Fig. 6.5 that the equivalent subband filters are given by

$$\begin{aligned} f_0(n) &= \phi_S(n) \\ f_k(n) &= \psi_{S+1-k}(n), \quad k = 1, 2, \ldots, S \end{aligned} \tag{6.17}$$

Furthermore, calling L_k the length of the kth subband filter, and recalling that $f_{\text{LP}}(n)$ and $f_{\text{HP}}(n)$ have lengths equal to $2N$, we see that (6.16) and (6.17) lead to

$$\begin{aligned} L_0 &= (2^S - 1)(2N - 1) + 1 \\ L_k &= (2^{S+1-k} - 1)(2N - 1) + 1, \quad k = 1, 2, \ldots, S \end{aligned} \tag{6.18}$$

In the frequency domain, (6.17) can be written as

$$\begin{aligned} F_0(e^{j\omega}) &= \prod_{i=0}^{S-1} F_{\text{LP}}(e^{j2^i\omega}) \\ F_k(e^{j\omega}) &= F_{\text{HP}}(e^{j2^{S-k}\omega}) \prod_{i=0}^{S-k-1} F_{\text{LP}}(e^{j2^i\omega}), \quad k = 1, 2, \ldots, S-1 \\ F_S(e^{j\omega}) &= F_{\text{HP}}(e^{j\omega}) \end{aligned} \tag{6.19}$$

Because the stages of the HLT are lapped transforms, the corresponding analysis filters are identical to the time-reversed synthesis filters, that is,

$$h_k(n) = f_k(L_k - 1 - n) \tag{6.20}$$

Therefore, the analysis filters have the same magnitude frequency responses of the synthesis filters, i.e., $|H_k(e^{j\omega})| = |F_k(e^{j\omega})|$.

6.3.3 Wavelet Transforms with the HLT

The wavelet concept came from multiresolution signal analysis in seismic signal processing, where it was suggested [7] that the basis functions $f_k(n)$ should have a constant shape. This idea was explored in the context of image coding, but with nonorthogonal basis functions, in [8]. The basis function at the finest time resolution, $f_S(n)$, is the *wavelet* $\psi_1(n)$, whereas the other $f_2(n)$ to $f_{S-1}(n)$ are dilated versions of the wavelet. The idea of a multiresolution transformation with basis functions that are dilated versions of a wavelet was analyzed in detail in [9], in a continuous-time framework. In [5], it was shown that it is possible to construct a set of finite-length orthogonal basis functions for a wavelet transform. In [10] a wavelet construction based on the ELT was presented.

The application of the wavelet decomposition to discrete-time signals was analyzed in [11], where the connection with filter banks was first noted. In [11] it was shown that a discrete wavelet transform has precisely the form of the HLT in (6.13) and (6.14), with an implementation identical to that in Fig. 6.5. The recursion in (6.16) leads to subband filters $f_1(n)$ to $f_S(n)$ that have approximately constant shapes, which correspond to dilated versions of the wavelet $\psi_1(n)$.

In [11] it was also pointed out that a wavelet basis composed of only dilated versions of a wavelet $\psi_1(n)$ is complete only when the number of functions is increased to infinity, that is, when the frequency resolution is made arbitrarily fine. This corresponds to $S \to \infty$ in the tree structure of the HLT. In practice, we must stop the dilation process at some resolution, by picking up a tree depth S such that the subbands number 0 and 1 have block sizes large enough to cover the long-duration features that may be present in the signal.

When we stop the subsampling process, there will be a remaining subspace of the signal that is not spanned by the dilated wavelets. This subspace corresponds precisely to the low-pass band, which is covered by the low-pass filter $f_0(n)$. Note that $f_0(n)$ is approximately a dilated version of the *scaling function* $\phi_1(n)$, which is the basic low-pass filter $f_{LP}(n)$ of each stage of the HLT. The denomination scaling function comes from the fact that the dilated versions of the wavelet at different scales are obtained by repeated $1:2$ interpolations and convolutions with $\phi_1(n)$, according to (6.16).

The infinite basis composed by all the filters $f_k(n + rR_k)$, with $k = 0, 1, \ldots, S$ and r an arbitrary integer, can be an orthogonal basis for the space of all infinite duration signals $x(n)$, if and only if the corresponding nonuniform filter bank has perfect reconstruction with lossless polyphase component matrices. Since the HLT is a lossless PR filter bank, we conclude that the HLT performs an orthogonal wavelet transform, with the advantage of a fast implementation.

6.3.4   Regularity

We know that the HLT implements an orthogonal multiresolution transform, as long as the polyphase component matrix $\mathbf{A}(z)$ in Fig. 6.5 is lossless. In order to characterize the HLT as a wavelet transform, though, it is necessary that the filters generated by the recursion in (6.16) be approximately dilated version of the wavelet $\psi_1(n)$. Due to the interpolation and convolution nature of (6.16), we see that a certain degree of smoothness in the scaling function $\phi_1(n)$ is required.

This problem was analyzed in great mathematical detail in [5]. One way to look at it is to consider the frequency response of the low-pass filter $f_0(n)$, which is given by (6.19), and can be rewritten as

$$F_0(e^{j\omega}) = \prod_{i=0}^{S-1} \Phi_1(e^{j2^i\omega}) \qquad (6.21)$$

where $\Phi_1(e^{j\omega})$ is the frequency response associated with the scaling function $\phi_1(n)$. The basic question is: given a $\Phi_1(e^{j\omega})$ that corresponds to a lossless $\mathbf{A}(z)$, will $F_0(e^{j\omega})$ be a good low-pass filter even as we increase the tree depth S?

In [5] it was shown that the product in (6.21) may converge to a discontinuous, fractal function, as $S \to \infty$. In order to avoid such a fractal behavior, i.e., to obtain a *regular* $F_0(e^{j\omega})$, it was shown in [5] that a sufficient condition is to put enough zeros of $\Phi_1(z)$ at $z = -1$. The maximum number of zeros that can be located at $z = -1$ is N [5]. [1] Therefore a maximally-regular (MR) choice of the low-pass filter of each stage (the scaling function) has the form

$$\Phi_1(z) = F_{\mathrm{LP}}(z) = (1 + z^{-1})^N U(z) \qquad (6.22)$$

where the zeros of $U(z)$ must be such that thelow-pass filter $f_{\mathrm{LP}}(n)$ and the high-pass filter $f_{\mathrm{HP}}(n) = (-1)^n f_{\mathrm{LP}}(2N - 1 - n)$ form a perfect reconstruction pair. It was shown in [5] that all solutions to the MR construction in (6.22) lead to the same $|\Phi_1(e^{j\omega})|$. Thus, all MR filters are spectral factors of the same half-band filter.

---

[1] We recall that $N - 1$ is the order of the polyphase component matrix $\mathbf{A}(z)$, and thus $\phi_1(n) = f_{\mathrm{LP}}(n)$ has length $2N$.

6.3.5   Examples of Wavelet Transforms

When we choose a set of PR filters $f_{LP}(n)$ and $f_{HP}(n)$ that do not satisfy (6.22) we will lose regularity. Thus, we can expect a loss of smoothness in $f_0(n)$ or $F_0(e^{j\omega})$. Fortunately, though, the loss of regularity is quite small as we move the zeros of $F_{LP}(z)$ away from $z = -1$. As long as $F_{LP}(z)$ is a good low-pass filter, with cutoff at $\pi/2$, $F_0(e^{j\omega})$ will probably also be a good low-pass filter, with cutoff at $\pi/2^S$.

In order to see this relation, we show in Figs. 6.9–6.12 examples of wavelet transforms for a tree with depth $S = 5$. In Fig. 6.9, we have chosen a memoryless $\mathbf{A}(z)$, in the form

$$\mathbf{A} = \sqrt{1/2} \begin{bmatrix} 1 & 1 \\ 1 & -1 \end{bmatrix} \tag{6.23}$$

which corresponds to the length-2 DCT matrix, according to (1.22). In this case, the HLT becomes a Haar transform [5]. Note that the maximum regularity condition of (6.22) is satisfied, because $F_{LP}(z) = \sqrt{1/2}(1 + z^{-1})$.

When we use a QMF filter bank for each stage of the hierarchical transform of Fig. 6.5, we do not get an orthogonal wavelet transform, because QMF filters are not strictly PR. However, since QMF filters lead to a complete aliasing cancellation, we can get quite close to an orthogonal wavelet transform if we use QMF filters with minimum reconstruction error, such as the ones in [12]. In Fig. 6.10 we have the basis functions generated from the QMF filters #12A in [12], for which $N = 6$.

In Fig. 6.11 we have the impulse and frequency responses of the MR orthogonal wavelet basis [5] for $N = 6$, and in Fig. 6.12 we have the responses of the orthogonal wavelet transform generated by the HLT structure of Fig. 6.5. Each stage of the HLT is an ELT with an overlapping factor $K = N/2 = 3$, and $\omega_s = 0.6\pi$.

We note that the smoothness of the basis functions $h_0(n)$ and $h_1(n)$ are about the same for all the wavelet transforms in Figs. 6.10 to 6.12. When we look at their frequency responses, their performances are roughly the same: the stopband attenuation is 27 dB for the QMF and MR wavelet bases, and 26 dB for the ELT-based HLT. The QMF-based wavelet transform has the disadvantage of not being orthogonal, and so it does not lead to perfect reconstruction.

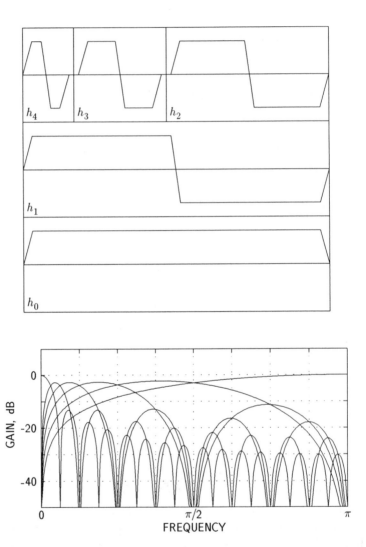

Figure 6.9. Top: first five basis functions of an orthogonal wavelet transform with $N = 1$, the Haar transform, for $S = 5$; bottom: the magnitude frequency responses of all basis functions. The frequency resolutions vary from $\pi/16$ to $\pi/2$. The samples of each $h_k(n)$ were connected by straight lines for easier viewing.

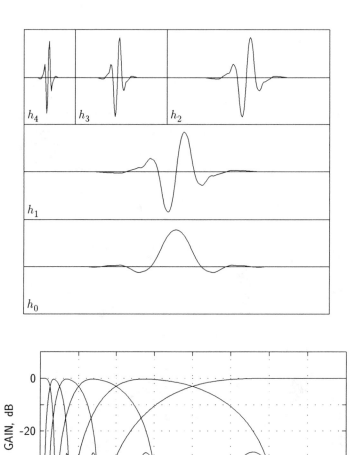

Figure 6.10. Top: first five basis functions of a quasi-orthogonal wavelet transform with $N = 6$ and $S = 5$, based on a QMF stage with Johnston #12A filters; bottom: the magnitude frequency responses of all basis functions. The frequency resolutions vary from $\pi/16$ to $\pi/2$.

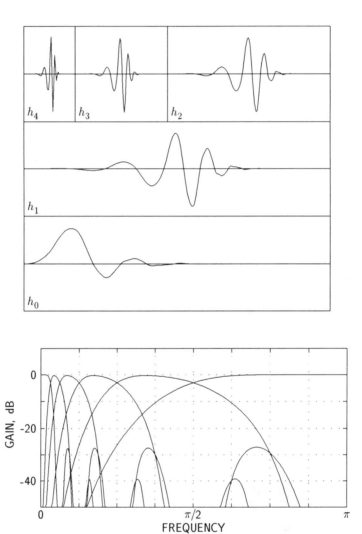

Figure 6.11. Top: first five basis functions of a maximally-regular (MR) orthogonal wavelet transform with $N = 6$ and $S = 5$, based on a PR two-channel stage with Daubechies filters D_6; bottom: the magnitude frequency responses of all basis functions. The frequency resolutions vary from $\pi/16$ to $\pi/2$.

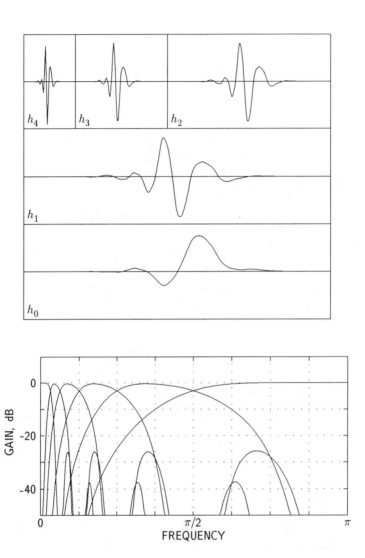

Figure 6.12. Top: first five basis functions of an orthogonal wavelet transform with $N = 6$ and $S = 5$, based on a hierarchical lapped transform (HLT), with each stage of the HLT being an ELT with $K = N/2 = 3$ and $\omega_s = 0.6\pi$; bottom: the magnitude frequency responses of all basis functions. The frequency resolutions vary from $\pi/16$ to $\pi/2$.

The QMF-based wavelet transform has basis functions with linear phase, whereas the maximally-regular (MR) and HLT wavelet bases do not have any symmetry. We recall from Chapter 3 that lossless two-channel PR filters cannot have linear phase. Therefore, an orthogonal wavelet bases cannot have symmetrical functions. In some applications, such as in image coding, it is easier to handle finite-length signals when the basis functions have linear phase. Therefore, some research has been devoted recently to the search for biorthogonal bases [13]. The basic idea is to remove the constraint in (6.20), that is, we allow the analysis and synthesis filters for the same band to be different. In this way, the added degrees of freedom allow the construction of PR nonuniform filter banks with symmetrical filters. The price to be paid for the filter symmetry is that the filtering performance is somewhat degraded [13]. Furthermore, the nonorthogonality of the bases generates some cross-correlation among the subband signals.

6.3.6   Computational Complexity

We saw in the examples in Figs. 6.10 to 6.12 that the ELT-based HLT leads to nonuniform filter banks with responses quite similar to those of the QMF and MR wavelet filter banks. However, the computational complexity of the HLT/ELT is significantly lower. The QMF filter bank can be computed with the polyphase decomposition of Fig. 3.15, the MR filter bank can be implemented with the lattice structures of Fig. 3.23, and the ELT-based HLT can be implemented with the fast ELT structures of Section 5.4. Then, the computational complexities of these wavelet transforms, in terms of number of operations per input sample, is

- QMF: N multiplications and N additions.

- MR: $N + 1$ multiplications and N additions.

- HLT/ELT: $(N + 3)/2$ multiplications and $(N + 3)/2$ additions.

The above complexities apply only to the first stage in Fig. 6.5. For the second stage, all the complexities are reduced by a factor of two, for the third they are reduced by a factor of four, and so on. With a large number of stages, the total complexity of the analysis or synthesis filter bank is approximately twice the values above.

For the examples of Figs. 6.10 to 6.12, where $N = 6$, we have the following complexities for the first stage:

WAVELET TRANSFORM	NUMBER OF MULTIPLICATIONS PER INPUT SAMPLE	NUMBER OF ADDITIONS PER INPUT SAMPLE	TOTAL
QMF	6	6	12
MR	7	6	13
ELT-BASED HLT	4.5	4.5	9

Therefore, we conclude that the HLT is a good choice for the implementation of a fast wavelet transform, since its complexity is lower than that of the maximally-regular wavelet transform. For larger values of N

6.4 Coding Performance

It is interesting to compare the transform coding gains G_{TC} [see Section 2.3] of the wavelet transforms in Figs. 6.9 to 6.12, as we did in Chapters 4 and 5 with the LOT, MLT, and ELT. Although G_{TC} may not predict directly the performance of a transform coder, differences in G_{TC} from one transform to another generally correlate well with differences in practical performance.

In Tables 6.1 to 6.3 we have the coding gains of the wavelet transforms based on the Haar transform, QMF filters, maximally-regular (MR) wavelets, and the ELT-based HLT. We have varied N from two to eight, in even values, except in the QMF case, where instead of using $N = 2$ we used $N = 1$ (note that the columns of the $\mathbf{A}$ matrix in (6.23) are the unique QMF filter bank with $N = 1$). For the HLT, we have used for each stage an ELT designed with $\omega_s = 0.6\pi$. The signal spectra were the same as in Chapters 4 and 5, that is, an AR(1) model and two measured speech spectra.

We can see from Tables 6.1 to 6.3 that the coding gains of all wavelet transforms are approximately the same, for a given value of N (the filters in each stage have lengths equal to $2N$). As expected, the gain increases slightly with N, except with only one stage and with the speech spectra. Again, the results reinforce our previous observation that the HLT has the same filtering and coding performance of the other wavelet transform, but the HLT has the advantage of a lower computational complexity.

With the AR(1) signal model, we become close to the maximum gain of 10.11 dB with five or six levels of resolution, and thus transform image coding based on

WAVELET	NUMBER OF STAGES, S					
TRANSFORM	1	2	3	4	5	6
QMF, $N = 1$ (HAAR)	5.06	7.19	7.96	8.20	8.27	8.28
QMF, $N = 4$	5.77	8.28	9.21	9.51	9.60	9.62
QMF, $N = 6$	5.87	8.44	9.40	9.70	9.77	9.78
QMF, $N = 8$	5.90	8.47	9.43	9.73	9.81	9.83
MR, $N = 2$	5.62	8.05	8.95	9.23	9.30	9.32
MR, $N = 4$	5.81	8.35	9.29	9.59	9.67	9.68
MR, $N = 6$	5.87	8.43	9.38	9.69	9.76	9.78
MR, $N = 8$	5.89	8.47	9.43	9.73	9.81	9.82
HLT, $K = N/2 = 1$	5.48	7.86	8.75	9.02	9.09	9.11
HLT, $K = N/2 = 2$	5.75	8.25	9.18	9.48	9.55	9.57
HLT, $K = N/2 = 3$	5.86	8.42	9.37	9.67	9.75	9.76
HLT, $K = N/2 = 4$	5.87	8.43	9.38	9.69	9.76	9.77

Table 6.1. Coding gain in dB of several wavelet transforms, for an AR(1) signal with $\rho = 0.95$.

WAVELET	NUMBER OF STAGES, S					
TRANSFORM	1	2	3	4	5	6
QMF, $N = 1$ (HAAR)	3.49	4.77	5.02	5.02	5.04	5.10
QMF, $N = 4$	3.92	6.20	6.74	6.74	6.76	6.89
QMF, $N = 6$	3.88	6.26	6.91	6.91	6.93	7.07
QMF, $N = 8$	3.85	6.26	6.99	6.99	7.00	7.15
MR, $N = 2$	3.96	5.89	6.30	6.30	6.32	6.43
MR, $N = 4$	3.92	6.22	6.78	6.78	6.80	6.94
MR, $N = 6$	3.88	6.26	6.91	6.91	6.92	7.07
MR, $N = 8$	3.85	6.26	6.98	6.98	7.00	7.14
HLT, $K = N/2 = 1$	3.89	5.78	6.17	6.17	6.19	6.30
HLT, $K = N/2 = 2$	3.95	6.16	6.65	6.65	6.67	6.81
HLT, $K = N/2 = 3$	3.89	6.25	6.89	6.89	6.90	7.05
HLT, $K = N/2 = 4$	3.88	6.26	6.90	6.90	6.92	7.06

Table 6.2. Coding gain in dB of several wavelet transforms, for a male speech signal.

Wavelet	Number of Stages, S					
Transform	1	2	3	4	5	6
QMF, $N = 1$ (Haar)	1.78	2.05	2.08	2.08	2.27	2.27
QMF, $N = 4$	2.87	3.19	3.25	3.25	3.62	3.63
QMF, $N = 6$	3.14	3.46	3.51	3.51	3.91	3.92
QMF, $N = 8$	3.25	3.56	3.61	3.61	4.01	4.02
MR, $N = 2$	2.45	2.76	2.81	2.81	3.12	3.12
MR, $N = 4$	2.94	3.26	3.30	3.30	3.69	3.70
MR, $N = 6$	3.13	3.45	3.50	3.50	3.90	3.90
MR, $N = 8$	3.24	3.55	3.59	3.60	4.00	4.01
HLT, $K = N/2 = 1$	2.41	2.72	2.76	2.76	3.06	3.06
HLT, $K = N/2 = 2$	2.77	3.09	3.13	3.13	3.50	3.51
HLT, $K = N/2 = 3$	3.10	3.42	3.46	3.46	3.86	3.87
HLT, $K = N/2 = 4$	3.13	3.44	3.49	3.49	3.89	3.89

Table 6.3. Coding gain in dB of several wavelet transforms, for a female speech signal.

wavelet transforms should give results as good as those obtained with block transforms [14]. However, with the speech spectra the coding gains are disappointing, when we compare with the uniform filter banks implemented by ELTs in Chapter 5. For the male speech spectrum, for example, the ELT can lead to coding gains higher than 12 dB, whereas with the wavelet transforms the coding gains do not get higher than 7.2 dB, even with more than six stages. A loss of 4.8 dB in SNR is significant enough to allow us to state that the wavelet transform, with its octave scale nonuniform subband splitting, may not be adequate for speech signals.

The main problem with speech signals is that we need fine frequency resolution not only at low-frequency subbands, but also at some of the middle-frequency subbands, in order to resolve the formant structure. With the generalized HLT of Fig. 6.7, we can obtain nonuniform subband divisions that are more adequate for speech coding. For example, in Table 6.4 we have the coding gains of an HLT based on stages with four subbands, with the two lowest-frequency subbands carried to further stages, i.e., $M = 4$ and $R = 2$.

In Table 6.4 we see that the coding gains of the generalized HLT are significantly higher than those of the wavelet transform. For example, with $S = 2$ and $K = 4$ in the generalized HLT, we obtain a coding gain for the male speech signal of 9.83 dB.

	AR(1), $\rho = 0.95$			MALE SPEECH			FEMALE SPEECH		
	S=1	S=2	S=3	S=1	S=2	S=3	S=1	S=2	S=3
N = 1	8.01	9.32	9.44	6.85	8.35	8.52	4.18	4.39	4.98
N = 2	8.40	9.80	9.93	7.69	9.30	9.51	4.44	4.67	5.38
N = 3	8.48	9.89	10.01	8.04	9.75	9.99	4.56	4.79	5.55
N = 4	8.50	9.90	10.03	8.10	9.83	10.09	4.58	4.81	5.58

Table 6.4. Coding gain in dB of the generalized HLT with $M = 4$ and $R = 2$.

For the wavelet HLT with $S = 4$ and $K = 4$, the coding gain for the same signal is only 6.90 dB. In both cases, the narrowest subband has the width $\pi/16$, but the improvement in coding gain with the generalized HLT is almost 3 dB.

6.5 Summary

The basic point that we have explored in this chapter was that multiresolution signal analysis and synthesis can be performed with a nonuniform filter bank. A simple way to generate nonuniform subband divisions is to use tree-structured filter banks,as we saw in Sections 6.1 and 6.2. The hierarchical lapped transform (HLT), which we discussed in Section 6.2, performs a nonuniform subband analysis and synthesis with perfect reconstruction. With the generalized HLT, we can generate a large variety of nonuniform subband divisions, as long as the subband decimation factors are highly composite integers.

When the HLT is based on a tree of two-band filters, an orthogonal wavelet transform is implemented, as we saw in Section 6.3. The HLT performs as well as other wavelet transforms, when we use an ELT to implement each two-band stage. However, the ELT-based HLT has the advantage of a lower computational complexity than other wavelet transforms.

The octave band-splitting that is generated with a wavelet transform is adequate for image coding, for example, since it leads not only to high transform coding gains, but it also allows for progressive transmission. However, for other signals, such as speech, we saw in Section 6.4 that the coding gains of the wavelet transforms were disappointingly low. With the generalized HLT, we can obtain better nonuniform subband divisions for speech coding. As we saw in Section 6.4, it is possible to construct a generalized HLT with the same scales of resolution as a wavelet transform, but with an improvement in coding gain of almost 3 dB.

References

[1] Lee, D. S., and K. H. Tzou, "Hierarchical DCT coding of HDTV for ATM networks," *IEEE Intl. Conf. Acoust., Speech, Signal Processing*, Albuquerque, NM, pp. 2249–2252, Apr. 1990.

[2] Wang, L., and M. Goldberg, "Reduced-difference pyramid: a data structure for progressive image transmission," *Opt. Eng.*, vol. 28, July 1989, pp. 708–716.

[3] Kishino, F., K. Manabe, Y. Hayashi, and H. Yasuda, "Variable bit-rate coding of video signals for ATM networks," *IEEE J. Selected Areas Commun.*, vol. 7, June 1989, pp. 801–806.

[4] Malvar, H. S., "Efficient signal coding with hierarquical lapped transforms," *IEEE Intl. Conf. Acoust., Speech, Signal Processing*, Albuquerque, NM, pp. 1519–1522, Apr. 1990.

[5] Daubechies, I., "Orthonormal bases of compactly supported wavelets," *Comm. Pure Appl. Math.*, vol. XLI, 1988, pp. 909–996.

[6] Evangelista, G., and C. W. Barnes, "Discrete-time wavelet transforms and their generalizations," *IEEE Intl. Symp. Circuits Syst.*, New Orleans, LA, pp. 2026–2029, May 1990.

[7] Morlet, J., G. Arens, E. Fourgeau, and D. Giard, "Wave propagation and sampling theory: parts I and II," *Geophysics*, vol. 47, Feb. 1982, pp. 203–236.

[8] Burt, P. J., and E. H. Adelson, "The Laplacian pyramid as a compact image code," *IEEE Trans. Commun.*, vol. COM-31, Apr. 1983, pp. 532–540.

[9] Grossman, A., and J. Morlet, "Decomposition of Hardy functions into square integrable wavelets of constant shape," *SIAM J. Math. Anal.*, vol. 47, July 1984, pp. 203–236.

[10] Coifman, R., and Y. Meyer, "Remarques sur l'analyse de Fourier à fenêtre," *C. R. Acad. Sci. Paris*, vol. 312, série I, 1991, pp. 259–261.

[11] Mallat, S. G., "A theory for multiresolution signal decomposition: the wavelet representation," *IEEE Trans. Pattern Anal. Machine Intell.*, vol. 11, July 1989, pp. 674–693.

[12] Johnston, J. D., "A filter family designed for use in quadrature mirror filter banks," *IEEE Intl. Conf. Acoust., Speech, Signal Processing*, Denver, CO, pp. 291–294, Apr. 1980.

[13] Vetterli, M., and C. Herley, "Wavelets and filter banks: theory and design," Tech. Rep. 206/90/36, CTR, Columbia University, New York, Aug. 1990.

[14] Mallat, S. G., "Multifrequency channel decompositions of images and wavelet models," *IEEE Trans. Acoust., Speech, Signal Processing*, vol. 37, Dec. 1989, pp. 2091–2110.

Chapter 7

Applications of Lapped Transforms

After studying in detail the basic theory of lapped transforms in the previous chapters, we will now present examples of the use of LTs in common signal processing problems. We do not intend to cover exhaustively all possible applications of LTs; our purpose is to highlight the influence of the transform on the signal processing system, and to compare the relative performance of LTs to those of traditional block transforms. For example, in our discussion of speech and image coding, we consider relatively simple coders; a discussion of the state-of-the-art in signal coding would go beyond the scope of this book.

In Chapter 2 we saw that the main areas of applications of block transforms were in spectrum estimation, signal filtering, fixed or adaptive, and signal coding. The use of LTs in spectrum estimation will be discussed in Section 7.1, and in Section 7.2 we will present examples of signal filtering with LTs. Most of the applications of signal coding deal either with speech or images, and the coding strategies are somewhat different in the two cases. Thus, we will discuss speech coding with LTs in Section 7.3, and the use of LTs in image coding, including progressive image transmission, in Section 7.4.

7.1 Spectrum Estimation

In Section 2.2 we saw that the power spectrum of a random signal can be estimated from an observation of the signal by means of averaged periodograms. In Fig. 2.9

247

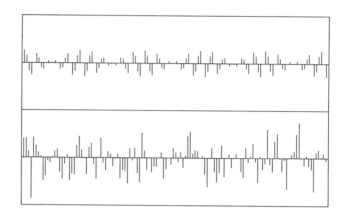

Figure 7.1. Waveforms for the spectrum estimation example. Top: signal composed of the sum of two sinusoids, at frequencies 0.36π and 0.44π; bottom: the same signal, corrupted with additive white Gaussian noise with mean-square amplitude 9 dB higher than the signal.

we noted that similar performances were obtained with several block transforms. We will now consider the use of a lapped transform instead of a block transform in the computation of periodograms.

Let us assume that our problem is to find two sinusoids in a noisy signal. This is a simple form of the most common problem in spectroscopy, where we want to find signal components buried in noise. In Fig. 7.1 we have a signal composed of the sum of two sinusoids whose frequencies are somewhat near: 0.36π and 0.44π. To that signal we have added white Gaussian noise, with a standard deviation 2.8 times higher than the rms amplitude of the original signal; this corresponds to a signal-to-noise ratio of -9 dB.

We have estimated the spectrum of the noisy signal by means of the averaged periodogram technique described in Section 2.2. In order to provide adequate noise filtering, we have taken the average of twenty periodograms. Each periodogram was computed either with the DFT, DCT, or ELT, with overlapping factors varying from $K = 1$ to $K = 4$. With the block transforms (DFT and DCT), we have windowed each signal block with a Hamming window [1]. The block size was varied from $M = 32$ to $M = 64$. Because a length-M DFT provides information on only $M/2$ frequencies, we have shifted each block by $M/2$ samples, when using the DFT.

The results of the spectrum estimation example are shown in Figs. 7.2 to 7.5, where the estimated power spectra are plotted. All spectra were normalized, with the spectral peak being set at the value of one. The vertical scale in all figures is $[0, 1.2]$.

With the DFT and $M = 32$, there is simply not enough resolution to capture the two sinusoids, and only one wide lobe appears in the spectrum. With $M = 64$, there is some indication of the existence of two sinusoids. For $M = 128$, the two frequency components are clearly distinguishable, well above the noise level. When we replace the DFT by the DCT, we see in Fig. 7.3 that the two sinusoids still cannot be resolved for $M = 32$, but they are clearly distinguishable for $M = 64$ and $M = 128$.

For the ELT, we see in Fig. 7.4 that even with $M = 32$ we can already distinguish the two sinusoids, mainly with larger values of the overlapping factor. For $M = 64$, Fig. 7.5 shows a performance similar to that of the DFT with $M = 128$ for all the values of the overlapping factor.

The better resolution of the ELT for a given block length M was expected, since we saw in Chapter 6 that the ELT basis functions are good bandpass filters, mainly for larger values of the overlapping factor K. In our example, reliable detection of the sinusoids could be done with any of the transforms, but with different block lengths. We could use a DFT with $M = 128$, a DCT with $M = 64$, or an ELT with $M = 32$ and $K = 2$ or $K = 4$. In the table below, we compare the computational complexities for each case. We have included the operations necessary for data windowing (DFT and DCT) and periodogram averaging. These operations amount to $2M$ multiplications and $M/2$ additions for the DFT, $2M$ multiplications and M additions for the DCT, and M multiplications and M additions for the ELT. We see that the DCT leads to about half of the computational complexity of the DFT. With the ELT, further reduction in the complexity is achieved (one third of the DFT complexity for $K = 2$).

TRANSFORM	MULTIPLICATIONS PER BLOCK	ADDITIONS PER BLOCK	TOTAL
DFT, $M = 128$	514	1092	1606
DCT, $M = 64$	321	577	898
ELT, $M = 32, K = 4$	288	416	704
ELT, $M = 32, K = 2$	224	352	576

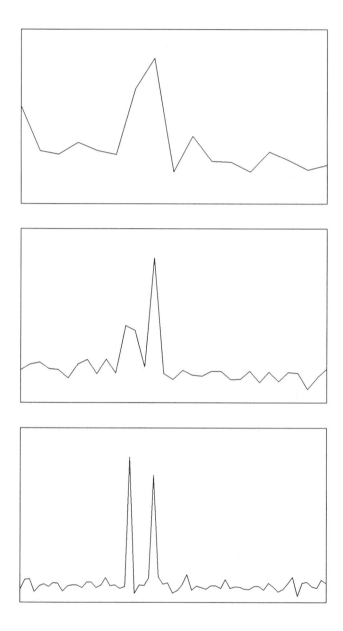

Figure 7.2. Spectrum estimates using the DFT, for $M = 32$ (top), $M = 64$ (middle), and $M = 128$ (bottom). Vertical scale is normalized squared magnitude. Horizontal frequency scale is $[0, \pi]$.

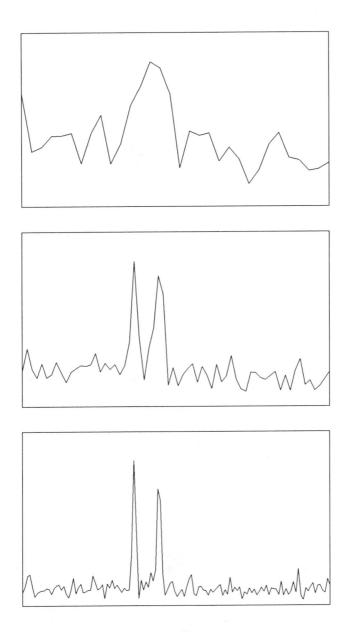

Figure 7.3. Spectrum estimates using the DCT, for $M = 32$ (top), $M = 64$ (middle), and $M = 128$ (bottom). Vertical scale is normalized squared magnitude. Horizontal frequency scale is $[0, \pi]$.

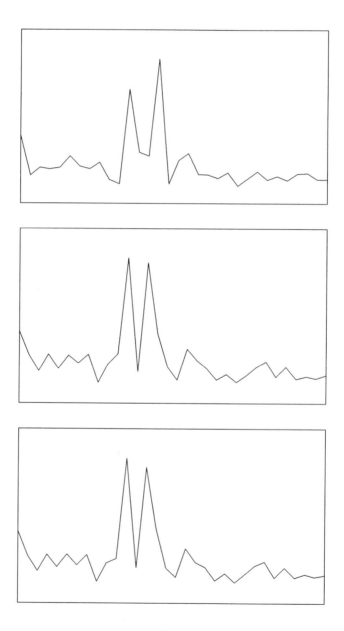

Figure 7.4. Spectrum estimates using the ELT, for $M = 32$. The overlapping factors are $K = 1$ (top), $K = 2$ (middle), and $K = 4$ (bottom). Vertical scale is normalized squared magnitude. Horizontal frequency scale is $[0, \pi]$.

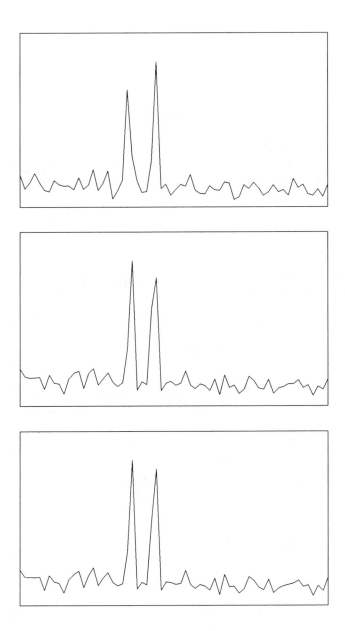

Figure 7.5. Spectrum estimates using the ELT, for $M = 64$. The overlapping factors are $K = 1$ (top), $K = 2$ (middle), and $K = 4$ (bottom). Vertical scale is normalized squared magnitude. Horizontal frequency scale is $[0, \pi]$.

Therefore, we conclude that the ELT can replace the DFT or DCT in some spectral estimation problems. This is particularly true when we want to resolve closely-spaced sinusoids, because of the better frequency resolution of the ELT.

7.2 Signal Filtering

Transforms can be used for efficient FIR filtering, fixed or adaptive, as we saw in Chapter 2. With lapped transforms, we can do a good job of signal filtering, without the blocking effects that are associated with traditional block transforms. In this section, we will consider the application of LTs to signal filtering.

7.2.1 FIR Filtering

We discussed in Section 2.1 that the DFT can be used for the efficient implementation of FIR filters, using the overlap-add or overlap-save procedures. These procedures are a direct consequence of the shift property of the DFT, which is not shared by other transforms. Therefore, exact shift-invariant FIR filtering with other transforms is not nearly as efficient as with the DFT.

Nevertheless, there are many applications, such as transform coding, where the signal will necessarily be subject to errors. Then, approximately shift-invariant filtering might be good enough. In a DCT-based image coder, for example, filtering in the DCT domain can lead to good results [2].

A block transform or a multirate filter bank can be used to perform signal filtering, as shown in Fig. 3.30. The basic idea is to do the simple processing of the subband signals:

$$\hat{X}_k(m) = c_k \, X_k(m)$$

where c_k is the gain for the kth subband. If we pick $c_k \equiv 1, \forall \ k$, then we recover $\hat{x}(n) = x(n - D)$ in Fig. 3.30, that is, no filtering is performed.

As we choose different values for the subband gains $c_0, \ldots, c_{M-1}$, we can reshape the signal spectrum in an appropriate way for the application at hand. In particular, if we set some of the c_k to zero, we can eliminate certain subbands from the signal, thereby achieving a filtering result.

Strictly speaking, we know from Section 3.6 that the perfect reconstruction property of block transforms or lapped transforms is lost if all c_k in Fig. 3.30 are not equal. The aliasing cancellation mechanism will not work, and the structure of Fig. 3.30 will actually be performing periodically time-varying filtering. With lapped transforms, though, the amount of aliasing is much smaller than with block transforms, due to the better bandpass responses of the basis functions of LTs. Thus, we can get satisfactory results with the structure of Fig. 3.30, as long as we use an LT.

In order to verify these ideas, let us consider a simple low-pass filtering example, in which we set the subband gains in Fig. 3.30 as

$$c_k = \begin{cases} 1, & k = 0, 1, \ldots, M/2 - 1 \\ 0, & k = M/2, M/2 + 1, \ldots, M - 1 \end{cases} \tag{7.1}$$

In this way, we do not modify the first $M/2$ subbands, but we remove the last $M/2$ subbands from the signal. The net effect is similar to low-pass filtering with a cutoff frequency of $\pi/2$.

We have processed a speech signal, sampled at 8 kHz, with the structure of Fig. 3.30, using the subband gains of (7.1), with $M = 32$. We have used the DCT, LOT, and ELT filter banks. As a standard of comparison, we have also filtered the same speech signal with a length-63 FIR low-pass filter with cutoff at $\pi/2$, designed with the window method and a Hamming window [1]. In all cases, the overall filtering response has linear phase because the impulse responses of the inverse transform are equal to the time-reversed impulse responses of the direct transform, as discussed in Chapter 3.

In Fig. 7.6 we show a segment of the processed speech signal for all cases. For easier viewing, we have connected subsequent samples of the signals by straight lines. The lack of aliasing cancellation with the DCT is manifested as blocking effects, as shown by the arrows in Fig. 7.6. With the lapped transforms, there are no visible blocking effects. Furthermore, when we listen to the processed signals, the blocking effects of the DCT-filtered signal are quite disturbing. The signals processed with the lapped transforms are virtually indistinguishable from the low-pass filtered signal.

From the results presented above, we conclude that it is possible to use the structure of Fig. 3.30 to perform signal filtering, in practice. If necessary, a variable filter can be easily implemented simply by varying the subband gains c_k.

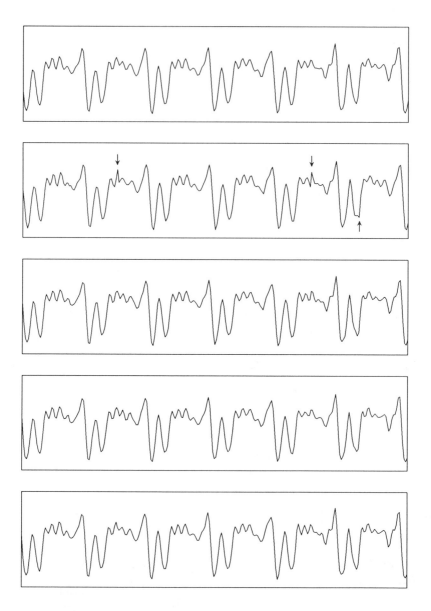

Figure 7.6. Waveforms resulting from low-pass filtering of a speech signal. Each waveform contains 256 samples. From top to bottom: FIR filter, block filter with the DCT, LOT, ELT with $K = 1$, and ELT with $K = 2$. Note the blocking effects with the DCT, pointed by the arrows.

The main advantage of LT-based filtering is in the computational complexity. For our example, the 63-point Hamming-windowed filter needs 32 multiplications and 63 additions per input sample, whereas the ELT-based filter with $K = 2$ needs 12.5 multiplications and 20 additions per input sample. Thus, the ELT leads to about one third of the computational complexity of the direct-form FIR filter.

In a transform coder, this filtering capability of LTs can be used to recover a low-pass filtered version of the signal, for example. This idea has been used recently in a compatible high-definition TV codec (coder-decoder) proposed by NTT in Japan [3].

7.2.2 Adaptive Filtering

adaptive In Chapter 2 we discussed how transforms could be used to improve the performance of adaptive filters. This is particularly true when the input signal to the adaptive filter has a low spectral flatness, i.e., when it is heavily colored. As we saw in Section 2.1, this leads to a large spread of eigenvalues of the input autocorrelation matrix. Such a spread slows down convergence of the adaptive filter [4].

We mentioned in Chapter 2 that one of the ways in which transforms can be used to improve the convergence rate of adaptive filters is with the type-I frequency-domain adaptive filter (FDAF-I) structure of Fig. 2.5 [5, 6]. If we replace the block transform **A** in Fig. 2.5 by a lapped transform, we can expect similar performances, but without blocking effects, according to the results of the previous subsection.

In Fig. 2.5 the adaptive filter for each subband has order zero, since it is composed of a single coefficient, but we can use higher order filters. In this way, the frequency-domain adaptive filter could implement the equivalent of a much higher order adaptive transversal filter, since a delay of z^{-1} in a subband signal corresponds to a delay of z^{-M} in the baseband signal. This idea has been used in [7] for acoustic echo cancellation, where filters of order higher than one thousand may be required.

Inspired by the echo cancellation problem of [7], which is quite important in audio and video teleconferencing, we designed a simple experiment to evaluate the performance of lapped transforms in the FDAF-I structure. Referring to Fig. 7.7, we have generated a colored signal $x(n)$ by passing white noise through an all-pole filter

$$H(z) = \prod_{i=1}^{6} \frac{1}{(1 - p_i z^{-1})}$$

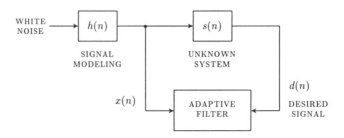

Figure 7.7. Generation of the input signal and the desired signal for the adaptive filtering example.

where the poles p_1 to p_6 were located at

$$p_{1,2} = 0.7e^{\pm j0.2\pi}$$
$$p_{3,4} = 0.8e^{\pm j0.3\pi}$$
$$p_{5,6} = 0.8e^{\pm j0.6\pi}$$

Echo estimation, as modeled in Fig. 7.7, is a system identification problem, in which the adaptive filter response should approximate that of the unknown system. The output of the adaptive filter, which would be an estimate of the echo-corrupted signal, could be subtracted from the received signal in order to achieve echo cancellation.

In our experiment, the unknown system was a length-20 FIR filter $s(n)$ with two strong impulses, representing echoes: $s(8) = -0.9$ and $s(16) = 0.8$. Smaller values of $s(n)$ were located near the two impulses, in order to simulate multiple reflections. A plot of $s(n)$ is shown in Fig. 7.8. The desired signal $d(n)$ is the convolution of $x(n)$ and $s(n)$. For the adaptive filter, we have used a transversal filter with 20 taps, and the FDAF-I structure of Fig. 2.5 with the DCT, LOT, and ELT with $K = 1$. In all FDAF structures, we have chosen $M = 4$, and the adaptive filter for each subband had order 5.

The adaptation algorithm in all cases was the normalized LMS [4], with automatic adjustment of the convergence factor μ by means of a simple variance estimator. For the FDAF-I structures, each subband had an independent adaptive filter, and each of them worked as described below.

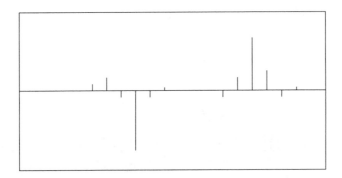

Figure 7.8. The reference system response for the adaptive filtering example.

Calling $w_i(n)$ the value of the ith adaptive filter coefficient at the nth time instant, we used the LMS update rule

$$w_i(n+1) = w_i(n) + 2\mu(n)\,e(n)\,x(n-i)$$

where the time-varying convergence factor $\mu(n)$ was given by

$$\mu(n) = \mu/\hat{\sigma}_x^2(n)$$

where $\hat{\sigma}_x^2(n)$ is an estimate of the input signal variance. We have used the simple recursive estimate

$$\hat{\sigma}_x^2(n+1) = \rho\,\hat{\sigma}_x^2(n) + (1-\rho)\,x^2(n)$$

The integrator gain ρ was set to 0.92, which gave a good compromise between noise level and speed of $\mu(n)$ adaptation. In each case, we have chosen for μ a value 30% lower than the upper limit for instability. For the FDAF filters, the optimum values of μ for each subband were different.

In Figs. 7.9 and 7.10 we have plots of the mean-square error level $E\{e^2(n)\}$ in dB as a function of adaptation time. Each curve is actually the average of 200 runs of the adaptive filter of Fig. 7.7. In Fig. 7.9 we have a zoom of the first 800 iterations, whereas Fig. 7.10 shows 4,000 iterations.

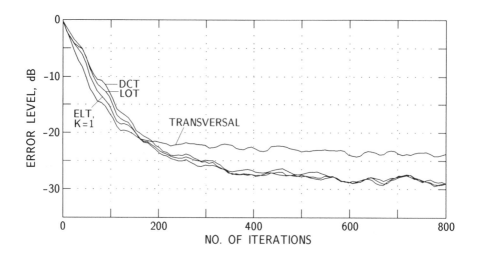

Figure 7.9. Learning curves of the several adaptive filters.

As expected, the LMS learning curve first starts reasonably fast with an initially small time constant. Then, the low eigenvalues of the input autocorrelation matrix dominate, and a much larger time constant enters into action [4, Fig. 6.5]. We should note that the error level will eventually reach values below −40 dB because the system to be identified is exactly FIR.

The learning curves of the FDAF-I filters are all quite similar, showing convergence to an error level close to −28 dB in about 200 iterations. The error stabilizes at this level because of the uncanceled aliasing that is present in the FDAF-I structure. In practice, a 28 dB reduction in the echo is probably good enough. The main advantage of using the FDAF-I adaptive filters is that their convergence time is roughly 20 times faster than that of the transversal filter.

Furthermore, the computational complexity of the transversal filter is approximately 40 operations per input sample, whereas the LOT-based FDAF filter requires 22 operations per input sample. In actual acoustic echo estimation problems, where the length of $s(n)$ may be above one thousand, the savings in computational complexity may be more than an order of magnitude [7].

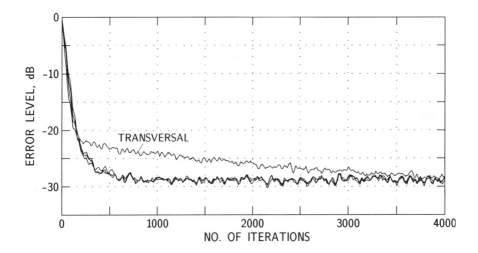

Figure 7.10. Learning curves of the several adaptive filters, with an expanded time scale.

Although all transforms have similar learning curves, we recall from the previous subsection that lapped transforms do not have blocking effects. So, LTs should be preferred over block transforms in FDAF-I adaptive filters for audio applications because the ear is quite sensitive to blocking effects.

7.3 Speech Coding

Signal coding can be performed in a transform domain, as we discussed in Section 2.3. In speech coding applications, transforms can be used whenever the signal delay is not a major concern. This is the case in voice mail, multimedia systems, audio response units, and other applications.

The basic block diagram of an adaptive transform coder that can be applied to speech is that of Fig. 2.12. The block transform **A** can be replaced by a lapped transform or another filter bank. In this way, the system of Fig. 2.12 can also be viewed as representing an adaptive subband coder.

If we use a lapped transform for adaptive speech coding with the system of Fig. 2.12, we can expect two basic improvements over block transforms:

- Absence of blocking effects.

- Higher signal-to-noise ratios.

We should not expect blocking effects with lapped transforms in view of our signal filtering experiment of the previous section. As for the signal-to-noise ratio, we recall from Chapters 4 and 5 that lapped transforms have higher coding gains than block transforms.

We have processed several speech signals with the adaptive transform coder of Fig. 2.12, using the DCT, LOT, and the ELT with overlapping factors $K = 1$ and $K = 4$. The transform length was chosen to be $M = 32$ in all cases. Each scalar quantizer Q_k was backwards adaptive (AQB) [8], so that no side information needed to be transmitted to the decoder. The bit allocation for each quantizer was recomputed at every new block, similar to the time-varying bit allocation of [9]. The allocation was based on the log-variance rule that we discussed in Section 2.3. The variances used for the allocation were estimated from the quantizer step sizes, so that no side information on the variance distribution was needed. The bit rate was set at 20 kbps, which corresponds to 80 bits per block, or 2.5 bits per sample.

In Fig. 7.11 we have a plot of the segmental signal-to-noise ratio (SNR) versus block index, for an input speech signal with 32,000 samples. Each segmental SNR curve was obtained by computing the SNR in dB for each block, and running a 10-point moving average to filter the curve.

The average segmental SNR was 13.4 dB for the DCT, 14.3 dB for the LOT, 17.2 dB for the ELT with $K = 1$, and 17.9 dB for the ELT with $K = 4$. Thus, the ELTs gave an improvement of about 4 dB in SNR over the DCT. Such an improvement is high enough to be clearly noticed. In listening tests, the average perceptible distortion level was noticeably lower with the lapped transforms[1]. Furthermore, the DCT-coded speech had a disturbing level of blocking effects.

An example of the blocking effect generation with the DCT is shown in Fig. 7.12. The reconstructed signals with the LOT or ELT do not suffer from this problem. It is also clearly seen in Fig. 7.12 that the reconstructed waveforms obtained with the LOT or ELT are much closer to the original than the reconstructed signal in the DCT-based coder.

---

[1] We would need a formal listening evaluation in order to better quantify the improvement of LTs over the DCT.

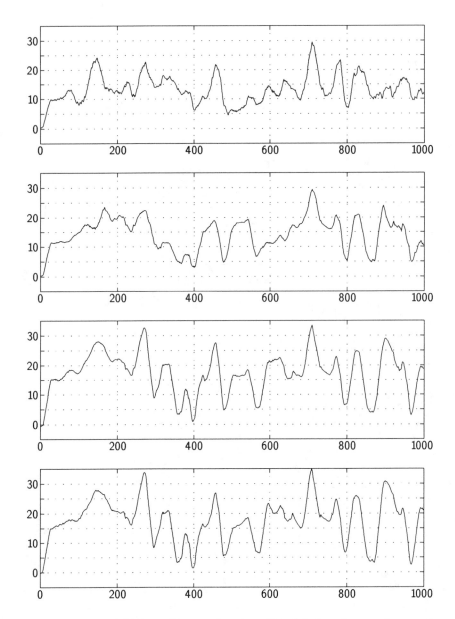

Figure 7.11. Segmental SNR, in dB, as a function of block index, for transform coding of speech at 20 kbps. The transform used in each case was, from top to bottom: DCT, LOT, ELT with $K = 1$, and ELT with $K = 4$.

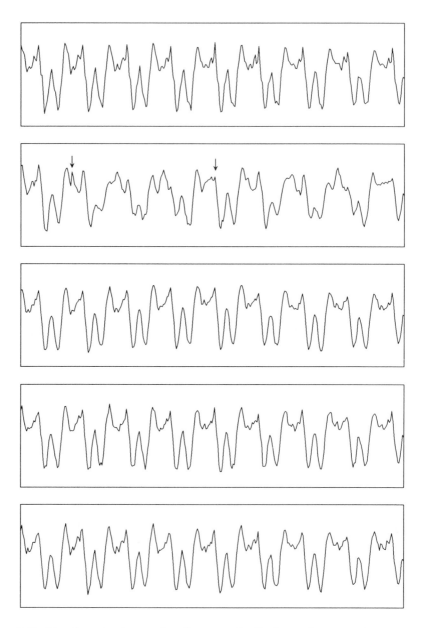

Figure 7.12. Waveforms in the speech coding example. Each waveform contains 332 samples, and corresponds to, from top to bottom: original signal, coded with the DCT, LOT, ELT with $K = 1$, and ELT with $K = 4$. Note the blocking effects with the DCT-coded speech, pointed by the arrows.

We note from Fig. 7.12 that the reconstructed waveform for the coder using the ELT with $K = 4$ is noticeably closer to the original than that of the ELT with $K = 1$. Thus, we conclude that larger values for the overlapping factor are useful in speech coding.

It is important to emphasize that the coder that we have used in our experiment is relatively simple, being recommended for medium to high rate applications where computational complexity must be kept low. In traditional transform coding of speech, it is usual to employ larger transform lengths ($M = 128$ or more). More elaborate spectrum estimation techniques are also used, in order to model the pitch periodicity of voiced speech [10]. With such coders, it is possible to reduce the bit rate down to 10 kbps, with very good speech quality. An extensive study of the use of lapped transforms in such coders has been carried out in [11], where it was shown that LTs lead to a virtually complete elimination of the blocking effects.

From the results in this section, it is clear that the ELT should be preferred over the LOT for speech coding because the ELT will lead to significantly higher SNRs in most cases. Furthermore, the use of overlapping factors of two or more may pay off in terms of SNR. However, the use of larger values of K leads to increased processing delays. Therefore, as we have already pointed out, transform coding of speech should be used only when the processing delay can be tolerated. An interesting example is in videophone applications, where the audio generally must be delayed to compensate for the video coding delay.

7.4 Image Coding

The basic motivation behind the development of lapped transforms was the reduction of blocking effects in image coding, as we noted in Section 1.3. We have performed a series of examples of adaptive transform coding (ATC) of images, using the DCT and LTs.

We have used the same basic ATC block diagram of Fig. 2.12. For the adaptive bit allocation rule, which corresponds to the box labeled "spectrum estimation" in Fig. 2.12, we have used the well-known Chen-Smith block classification rule [12]. The details of the bit allocation are described next.

Chen-Smith Adaptive Bit Allocation Algorithm

Step 1: Starting with an image of size $N \times N$ pixels, compute the transform of each $M \times M$ block.

Step 2: For the (i,j)th transformed block, $X_{ij}(n,m)$ compute its ac energy by

$$E_{AC}(i,j) = \sum_{n=0}^{M-1} \sum_{m=0}^{M-1} X_{ij}^2(n,m) - X_{ij}^2(0,0)$$

Step 3: Assign each block to one of the Q classes, according to

$$t_r \leq E_{AC}(i,j) \leq t_{r+1} \quad \Longrightarrow \quad c(i,j) = r$$

where $c(i,j)$ is the class index for block (i,j). The thresholds t_0 and t_Q are defined by

$$t_0 \equiv \min E_{AC}(i,j), \qquad t_Q \equiv \max E_{AC}(i,j)$$

and the other thresholds t_1 to t_{Q-1} are chosen such that all classes are equally populated.

Step 4: Estimate the variances of the transform coefficients for each class r by

$$\hat{\sigma}_r^2(n,m) = \frac{1}{R} \sum_{i,j \,|\, c(i,j)=r} \{X_{ij}(n,m) - E[X_{ij}(n,m)]\}^2$$

where R is the number of blocks in each class, which is given by

$$R = \frac{N^2}{Q\,M^2}$$

and $E[X_{ij}(n,m)]$ is estimated by running a simple average over all $X_{ij}(n,m)$ such that $c(i,j) = r$.

Step 5: Given all variances $\hat{\sigma}_r^2(n,m)$ for class r, define the number of bits for each transform coefficient, $b_r(n,m)$, according to the log-variance rule of Section 2.3. For each class r, the side information to be transmitted to the decoder is composed of $\hat{\sigma}_r^2(0,0)$ and $b_r(n,m), n,m = 0,1,\ldots,M-1$.

It is important to note that the Chen-Smith adaptive image coder is quite efficient, but there are certainly other coders that will outperform it. An example is the JPEG system [13], which is to be adopted as an image coding standard by the CCITT. A state-of-the art study of transform/subband coding of images and video can be found in [14]. As was the case with the speech coding example, our choice of the C&S algorithm was based on the idea of picking up a simple but efficient coding system, where we can highlight the influence of the transform on the overall performance.

Once the transform coefficient energies and bit allocations are known, the coefficients are quantized and sent to the decoder, according to Fig. 2.12. At the decoder, each transform coefficient is reconstructed by inverse quantization, with the bit patterns that were transmitted as side information. Then, each block is inverse transformed and the image is reconstructed.

In our experiments we have used 256×256 images, i.e. $N = 256$, and a block size $M = 8$. The number of classes was chosen as $Q = 4$, which is a good compromise between adaptability versus amount of side information [12]. In all cases, the original image was the standard "Lena" image, shown in Fig. 7.13.

The transforms that we used in our experiment were the DCT, LOT, MLT (ELT with $K = 1$), ELT with $K = 2$, and an HLT based on the structure of Fig. 6.5, with a tree depth of $S = 3$ and ELTs with $K = 1$ at each stage. All transforms were separable, that is, we transformed first the rows and then the columns of the image, as discussed in Section 1.2. For the lapped transforms, the computation at the borders of the image were done as described in Chapters 4 and 5.

Instead of just coding the image at a given bit rate, we have simulated a progressive transmission system [15]. The Chen-Smith adaptive bit allocation algorithm was used to determine the bit patterns for an average rate of one bit per pixel. Then, we have reconstructed the image at rates of 0.2, 0.4, 0.6, and 0.8 bits per pixel, including overhead. The partial image reconstruction was obtained by transmitting first the coefficients with highest energy for each class. Note that the decoder knows in what order the coefficients should be transmitted, since it has received the bit patterns and energies for all classes in the first step of the progressive transmission.

The results of our image coding experiments are shown in Figs. 7.14 to 7.17. In terms of signal-to-noise ratio, the LOT gave an average improvement of about

Figure 7.13. Original "Lena" image, 256×256 pixels, 8 bits per pixel.

0.4 dB over the DCT, at each progressive transmission stage. Such an improvement was predicted by the transform coding gains that we presented in Table 4.1 (we recall that the AR(1) model with $\rho = 0.95$ is adequate for most images [8]).

The main advantage of the LOT over the DCT is the reduction of blocking effects. At the initial rate of 0.2 bits/pixel, only a few low-frequency coefficients are transmitted per block, so that some blocking still appears with the LOT. At the higher rates, the blockiness in the LOT-coded images is less apparent than with the DCT-coded ones. For example, at 0.8 bits/pixel, there are some blocking effects near the shoulder in the DCT-coded image, whereas the LOT-coded image has no blocking effects.

For the MLT (ELT with $K = 1$), we obtained essentially the same image qualities as for the LOT. The SNRs were approximately the same, as well as the noticeable reduction in blocking effects. For $M = 8$, the LOT is computationally more efficient than the MLT, and so the LOT should be preferred over the MLT.

With the ELT ($K = 2$) the results were disappointing. According to the theory, we should expect a slight improvement in the SNR over the LOT, in the order of 0.3 dB [see Table 5.2]. However, the ELT-coded images had worse SNRs than the LOT-coded ones: 0.1 dB worse at 0.2 bits/pixel, and 0.8 dB worse at 0.8 bits/pixel. The problem is clearly in the borders of the image, which were not well reconstructed.

Figure 7.14. Progressive ATC transmission. Reconstructed images after 0.2 (top left), 0.4 (top right), 0.6 (bottom left), and 0.8 (bottom right) bits per pixel. Transform: DCT. Signal-to-noise ratios are 15.2 dB, 20.4 dB, 23.3 dB, and 25.1 dB.

Figure 7.15. Progressive ATC transmission. Reconstructed images after 0.2 (top left), 0.4 (top right), 0.6 (bottom left), and 0.8 (bottom right) bits per pixel. Transform: LOT. Signal-to-noise ratios are 15.9 dB, 20.8 dB, 23.7 dB, and 25.5 dB.

Figure 7.16. Progressive ATC transmission. Reconstructed images after 0.2 (top left), 0.4 (top right), 0.6 (bottom left), and 0.8 (bottom right) bits per pixel. Transform: ELT, $K = 2$. Signal-to-noise ratios are 15.8 dB, 20.6 dB, 23.2 dB, and 24.7 dB.

Figure 7.17. Progressive ATC transmission. Reconstructed images after 0.2 (top left), 0.4 (top right), 0.6 (bottom left), and 0.8 (bottom right) bits per pixel. Transform: HLT, $M = 2$, $K = 1$, $S = 3$. Signal-to-noise ratios are 16.1 dB, 19.9 dB, 22.2 dB, and 23.7 dB.

In order to account for the finite length of the image, we have used the structure of Fig. 5.12. Although that structure allows for perfect reconstruction, the basis functions for the first two and last two blocks of the signal are different than those of the middle blocks. Therefore, to code an image appropriately with the ELT, we should assign border blocks to different classes. The LOT does not suffer from this problem because its basis functions have linear phase.

Another problem with the ELT is the presence of ringing artifacts, which are caused by the longer basis functions of the ELT. In speech coding, such artifacts are not important, but in images they are immediately detected by the eye. Therefore, it seems that the LOT is more appropriate for image coding, whereas the ELT should be preferred for speech coding.

When we used the HLT, lower SNRs were expected, since the coding gain of the HLT with $K = 1$ and $S = 3$ is 0.4 dB lower than that of the DCT for $M = 8$ [see Tables 5.1 and 6.1]. The SNRs in the HLT-coded images in Fig. 7.17 are not quite in agreement with this theoretical prediction. In fact, for the first stage of the progressive transmission, at 0.2 bits/pixel, the HLT-coded image has an SNR that is 0.9 dB above that of the DCT. As the rate increases, the HLT performance degrades, and at 0.8 bit/pixel the HLT-code image has an SNR that is 1.4 dB below that of the DCT-coded image.

The HLT-coded images show strong artifacts at the lower rates, but no blocking affects. As the rates increase, these artifacts disappear: at 0.8 bits/pixel there are no annoying artifacts, but the noise level is visibly higher than that of the DCT. In order to improve the HLT performance, we should increase S, the number of tree levels in the HLT. In fact, according to the results obtained in [16], we should use the largest possible value for S.

7.5 Other Applications

From the results in the previous sections, it is clear that lapped transforms can be used in any application where a block transform or filter bank was originally employed. Depending on the characteristics of the signal to be processed and the kind of processing to be performed, a different lapped transform may be more appropriate.

Applications where we expect a performance gain when we use a lapped transform instead of a block transform or another filter bank include speech scrambling and transmultiplexing. In the first case, it is because of the blocking effect reduction associated with LTs. In the second, it is due to the lower complexity of ELTs when compared to other filter banks.

7.6 Summary

In this chapter we have put lapped transforms to work in several applications: spectrum estimation, signal filtering — fixed and adaptive, and signal coding — specifically speech and image coding. In all cases we had a better performance of LTs when compared to standard block transforms such as the DCT.

In spectrum estimation, we saw in Section 7.1 that the better filtering performance of LTs, when compared to the FFT, led to an improved frequency resolution. An ELT for a given block size M generally leads to a better discrimination of spectral components than a DFT with block size $2M$.

We saw in Section 7.2 that LTs can be used to approximate shift-invariant filtering with a lower computational complexity than direct-form filtering, and without the blocking effects associated with block transforms. In adaptive filtering, we saw that LTs had learning curves similar to those of the DCT, but again the reduction of blocking effects favors the use of LTs.

The advantages of LTs over the DCT in speech coding were clearly seen in Section 7.3. We noted that the improvement in signal-to-noise ratio of the ELT over the DCT was as high as 4 dB. The use of higher overlapping factors such as $K = 4$ pays off in terms of lower coding error. Furthermore, the absence of blocking effects is another advantage of LTs. Although the ELT has a higher computational complexity than the DCT, the ELT-based speech coder with $M = 32$ that we described in Section 7.3 would certainly fit in a single fixed-point DSP processor chip.[2]

For image coding applications, it is clear from the result in Section 7.4 that the LOT is the most appropriate lapped transform. Larger overlapping factors lead to the appearance of ringing around edges. Furthermore, the asymmetrical basis functions of the ELT lead to border effects. With the LOT, we have two gains over

---

[2]The computation of the direct and inverse ELT with $K = 4$ would need 40 operations per input sample, whereas an inexpensive DSP processor chip could perform more than 500 operations per input sample.

the DCT: a reduction of the blocking effects and a slight increase (by about 0.5 dB) in the signal-to-noise ratio.

As a final note we could say that, for all the applications that we discussed in this chapter, the advantages of LTs over standard block transforms are certainly due to the better filtering performance associated with the basis functionsof LTs. Therefore, it is clear that similar improvements could be expected if other filter banks were used, such as the QMF filter bank. The key advantage of LTs is that they are filter banks with very fast algorithms.

References

[1] Oppenheim, A. V., and R. Schafer, *Discrete-Time Signal Processing*, Englewood Cliffs, NJ: Prentice-Hall, 1989.

[2] Chitpraset, B., and K. R. Rao, "Discrete cosine transform filtering," *IEEE Intl. Conf. Acoust., Speech, Signal Processing*, Albuquerque, NM, pp. 1281–1284, Apr. 1990.

[3] Jozawa, H., and H. Watanabe, "Intrafield/interframe adaptive lapped transform for compatible HDTV coding," *Fourth Int. Conf. HDTV and Beyond*, Torino, Italy, Sep. 1991.

[4] Widrow, B., and S. D. Stearns, *Adaptive Signal Processing*, Englewood Cliffs, NJ: Prentice-Hall, 1985.

[5] Dentino, M., J. McCool, and B. Widrow, "Adaptive filtering in the frequency domain," *Proc. IEEE*, vol. 66, Dec. 1978, pp. 1658–1659.

[6] Narayan, S., A. M. Peterson, and M. J. Narasimha, "Transform domain LMS algorithm," *IEEE Trans. Acoust., Speech, Signal Processing*, vol. ASSP-31, June 1983, pp. 609–615.

[7] Gilloire, A., and M. Vetterli, "Adaptive filtering in subbands," *IEEE Intl. Conf. Acoust., Speech, Signal Processing*, New York, pp. 1572–1575, Apr. 1988.

[8] Jayant, N. S., and P. Noll, *Digital Coding of Waveforms*, Englewood Cliffs, NJ: Prentice-Hall, 1984.

[9] Honda, M., and F. Itakura, "Bit allocation in time and frequency domains for predictive coding of speech," *IEEE Trans. Acoust., Speech, Signal Processing*, vol. ASSP-32, June 1984, pp. 465–473.

[10] Cox, R. V., and R. E. Crochiere, "Real-time simulation of adaptive transform coding," *IEEE Trans. Acoust., Speech, Signal Processing*, vol. ASSP-29, Apr. 1981, pp. 147–154.

[11] Duarte, R., *Codificação Adaptativa de Voz Utilizando Transformadas com Super-posição*, Master's thesis, Universidade de Brasília, Brasília, Brazil, June 1990.

[12] Chen, W. H., and C. H. Smith, "Adaptive coding of monochrome and color images," *IEEE Trans. Commun.*, vol. COM-25, Nov. 1977, pp. 1285–1292.

[13] Wallace, G. K., "The JPEG still picture compression standard," *Commun. ACM*, vol. 34, Apr. 1991, pp. 31–44.

[14] Husøy, J. H., *Subband Coding of Still Images and Video*, PhD thesis, The Norwegian Institute of Technology, Trondheim, Norway, Jan. 1991.

[15] Chitpraset, B., and K. R. Rao, "Human visual weighted progressive image transmission," *IEEE Trans. Commun.*, vol. 38, July 1990, pp. 1040–1044.

[16] Mallat, S. G., "Multifrequency channel decompositions of images and wavelet models," *IEEE Trans. Acoust., Speech, Signal Processing*, vol. 37, Dec. 1989, pp. 2091–2110.

Appendix A

Tables of CQF Filter Coefficients

In this Appendix we present two tables of conjugate quadrature filters (CQF). The tables were generated using the method described in Section 3.4, based on the design of a half-band filter $u(n)$ with length $2L - 1$, where L is the desired length of the CQF filter pair. The impulse responses of the low-pass filter $h_0(n)$ are given in Tables A.1 and A.2, for several values of L. We recall from Section 3.4 that the high-pass impulse response is given by

$$h_1(n) = (-1)^n h_0(L - 1 - n)$$

The half-band filter $u(n) = h_0(n) * h_0(-n)$ was designed using the window method and a Kaiser window with parameter β. For the filters in Table A.1, we have used $\beta = 5$, which leads to stopband attenuations close to 20 dB. The filters in Table A.2 were designed with $\beta = 10$, which leads to stopband attenuations around 40 dB.

As discussed in Section 3.4, for a given half-band filter $u(n)$ with zeros in reciprocal pairs, there are many possible spectral factors, and each factor is a PR low-pas CQF filter. For all the filters in Tables A.1 and A.2, we have picked up a mix of zeros inside and outside the unit circle, in order to centralize the main lobe of $h_0(n)$.

L=4	L=8	L=12	L=16	L=20	L=24	L=28	L=32
0.0543	0.1518	0.0875	0.0723	0.0496	0.0482	0.0380	0.0375
0.4786	0.6267	0.0976	0.1046	0.0366	0.0440	0.0203	0.0256
0.7944	0.7132	0.0969	-0.0078	0.0132	-0.0196	0.0009	-0.0175
0.3700	0.1217	0.3277	0.0301	0.0580	0.0138	0.0245	0.0090
	-0.2241	0.6943	0.4053	0.1491	0.1424	0.0706	0.0776
	-0.0444	0.4869	0.7311	0.1577	0.1638	0.0585	0.0695
	0.0904	-0.0876	0.4450	0.2355	0.0327	0.0575	-0.0073
	-0.0219	-0.3393	-0.1062	0.4498	0.0975	0.1125	0.0359
		-0.0811	-0.2450	0.5993	0.4613	0.1823	0.1802
		0.1272	-0.0635	0.2091	0.6826	0.1862	0.1975
		0.0270	0.0744	-0.3061	0.3100	0.2883	0.0611
		-0.0242	-0.0342	-0.3665	-0.2109	0.4658	0.1250
			-0.0368	-0.0252	-0.3116	0.4896	0.4606
			0.0339	0.1937	-0.1070	0.0255	0.6277
			0.0310	0.0310	0.0699	-0.4151	0.2262
			-0.0215	-0.1075	-0.0090	-0.3492	-0.2788
				-0.0257	-0.0297	0.0221	-0.3624
				0.0674	0.0204	0.1984	-0.1345
				0.0182	0.0282	0.0043	0.0738
				-0.0247	-0.0287	-0.1260	0.0088
					-0.0116	-0.0148	-0.0296
					0.0278	0.0911	0.0068
					0.0191	0.0137	0.0271
					-0.0210	-0.0574	-0.0213
						-0.0101	-0.0132
						0.0459	0.0207
						0.0121	0.0157
						-0.0227	-0.0212
							-0.0045
							0.0230
							0.0136
							-0.0200

Table A.1. Filter coefficients for $\beta = 5$.

L=4	L=8	L=12	L=16	L=20	L=24	L=28	L=32
0.0059	-0.0719	0.0374	-0.0200	0.0127	-0.0088	0.0065	-0.0050
0.4160	-0.0425	-0.0011	-0.0000	-0.0000	0.0000	-0.0000	-0.0000
0.8146	0.5624	-0.1831	0.0804	-0.0445	0.0281	-0.0194	0.0142
0.4043	0.7939	-0.0273	0.0094	-0.0037	0.0020	-0.0011	0.0008
	0.2023	0.5853	-0.2085	0.1075	-0.0638	0.0419	-0.0293
	-0.0735	0.7418	-0.0598	0.0221	-0.0094	0.0053	-0.0031
	-0.0009	0.2413	0.5326	-0.2146	0.1237	-0.0773	0.0532
	0.0003	-0.1050	0.7408	-0.0736	0.0347	-0.0149	0.0097
		-0.0443	0.2885	0.5064	-0.2110	0.1340	-0.0862
		0.0154	-0.1347	0.7328	-0.0775	0.0468	-0.0195
		0.0007	-0.1013	0.2989	0.4920	-0.2008	0.1418
		-0.0002	0.0346	-0.1606	0.7237	-0.0715	0.0598
			0.0249	-0.1400	0.2946	0.4875	-0.1866
			-0.0069	0.0503	-0.1855	0.7134	-0.0622
			-0.0009	0.0555	-0.1675	0.2795	0.4853
			0.0002	-0.0164	0.0632	-0.2121	0.7018
				-0.0167	0.0811	-0.1886	0.2592
				0.0039	-0.0258	0.0740	-0.2403
				0.0010	-0.0363	0.1006	-0.2081
				-0.0002	0.0093	-0.0349	0.0820
					0.0125	-0.0545	0.1158
					-0.0025	0.0154	-0.0430
					-0.0011	0.0262	-0.0705
					0.0002	-0.0060	0.0212
						-0.0101	0.0399
						0.0018	-0.0100
						0.0012	-0.0203
						-0.0002	0.0041
							0.0086
							-0.0013
							-0.0013
							0.0002

Table A.2. Filter coefficients for $\beta = 10$.

Appendix B

Programs for Block Transforms

The purpose of this Appendix is to present computer programs that implement the fast algorithms for block transforms discussed in Section 2.5, namely the DFT, DHT, DCT, and DCT-IV. All programs were written in ANSI-standard "C", and they should be able to run unmodified in a wide variety of machines. They have been extensively tested, and should produce correct results. However, the author does not warrant that the programs are absolutely free from errors or that they will meet any specific requirements. The responsibility for the use of information obtained from those programs lies entirely on the user.

Although the programs were written with particular care to efficiency, the main goal was to make them fully recursive. Therefore, the reader should be careful to ensure that the available "C" compiler and linker will reserve enough stack space. In all systems where we have tested the programs the default stack sizes were enough for transform lengths up to $M = 1024$.

In all programs, complex multiplications and real orthogonal butterflies were implemented using the three-multiply/three-add procedure described in Section 2.5. In this way, the arithmetic operation counts for the programs are precisely those is Tables 2.1 and 2.2. Most of the data unshuffling steps were moved outside the recursive modules, so that all the reordering could be performed only once.

```
/*------------------------------------------------------------------------*
 *      SRFFT.C  -  Split-Radix Fast Fourier Transform                    *
 *                                                                        *
 *      This is a decimation-in-frequency version, using the steps        *
 *      of the algorithm as described in Section 2.5.                     *
 *                                                                        *
 *      Author:  Henrique S. Malvar.                                      *
 *      Date:    October 8, 1991.                                         *
 *                                                                        *
 *      Usage:   srfft(xr, xi, logm);   --  for direct DFT                *
 *               srifft(xr, xi, logm);  --  for inverse DFT               *
 *                                                                        *
 *      Arguments:  xr (float)  - input and output vector, length M,      *
 *                                real part.                              *
 *                  xi (float)  - imaginary part.                         *
 *                  logm (int)  - log (base 2) of vector length M,        *
 *                                e.g., for M = 256 -> logm = 8.          *
 *------------------------------------------------------------------------*/

#include <stdio.h>
#include <stdlib.h>
#include <alloc.h>
#include <math.h>

#define  MAXLOGM    11    /* max FFT length = 2^MAXLOGM */
#define  TWOPI      6.28318530717958647692
#define  SQHALF     0.707106781186547524401

/*------------------------------------------------------------------------*
 *      Error exit for program abortion.                                  *
 *------------------------------------------------------------------------*/

static   void   error_exit(void)
{
   exit(1);
}


/*------------------------------------------------------------------------*
 *      Data unshuffling according to bit-reversed indexing.              *
 *                                                                        *
 *      Bit reversal is done using Evan's algorithm (Ref: D. M. W.        *
 *      Evans, "An improved digit-reversal permutation algorithm ...",    *
 *      IEEE Trans. ASSP, Aug. 1987, pp. 1120-1125).                      *
 *------------------------------------------------------------------------*/

static   int   brseed[256];   /* Evans' seed table */
static   int   brsflg;        /* flag for table building */
```

```
void BR_permute(float *x, int logm)
{
    int      i, j, imax, lg2, n;
    int      off, fj, gno, *brp;
    float    tmp, *xp, *xq;

    lg2 = logm >> 1;
    n = 1 << lg2;
    if (logm & 1) lg2++;

    /* Create seed table if not yet built */
    if (brsflg != logm) {
        brsflg = logm;
        brseed[0] = 0;
        brseed[1] = 1;
        for (j = 2; j <= lg2; j++) {
            imax = 1 << (j - 1);
            for (i = 0; i < imax; i++) {
                brseed[i] <<= 1;
                brseed[i + imax] = brseed[i] + 1;
            }
        }
    }

    /* Unshuffling loop */
    for (off = 1; off < n; off++) {
        fj = n * brseed[off]; i = off; j = fj;
        tmp = x[i]; x[i] = x[j]; x[j] = tmp;
        xp = &x[i];
        brp = &brseed[1];
        for (gno = 1; gno < brseed[off]; gno++) {
            xp += n;
            j = fj + *brp++;
            xq = x + j;
            tmp = *xp; *xp = *xq; *xq = tmp;
        }
    }
}
/*-------------------------------------------------------------------*
 *      Recursive part of the SRFFT algorithm.                       *
 *-------------------------------------------------------------------*/

void srrec(float *xr, float *xi, int logm)
{
    static   int      m, m2, m4, m8, nel, n;
    static   float    *xr1, *xr2, *xi1, *xi2;
    static   float    *cn, *spcn, *smcn, *c3n, *spc3n, *smc3n;
    static   float    tmp1, tmp2, ang, c, s;
    static   float    *tab[MAXLOGM];
```

```c
/* Check range of logm */
if ((logm < 0) || (logm > MAXLOGM)) {
   printf("Error : SRFFT : logm = %d is out of bounds [%d, %d]\n",
      logm, 0, MAXLOGM);
   error_exit();
}

/* Compute trivial cases */
if (logm < 3) {
   if (logm == 2) {  /* length m = 4 */
      xr2  = xr + 2;
      xi2  = xi + 2;
      tmp1 = *xr + *xr2;
      *xr2 = *xr - *xr2;
      *xr  = tmp1;
      tmp1 = *xi + *xi2;
      *xi2 = *xi - *xi2;
      *xi  = tmp1;
      xr1  = xr + 1;
      xi1  = xi + 1;
      xr2++;
      xi2++;
      tmp1 = *xr1 + *xr2;
      *xr2 = *xr1 - *xr2;
      *xr1 = tmp1;
      tmp1 = *xi1 + *xi2;
      *xi2 = *xi1 - *xi2;
      *xi1 = tmp1;
      xr2  = xr + 1;
      xi2  = xi + 1;
      tmp1 = *xr + *xr2;
      *xr2 = *xr - *xr2;
      *xr  = tmp1;
      tmp1 = *xi + *xi2;
      *xi2 = *xi - *xi2;
      *xi  = tmp1;
      xr1  = xr + 2;
      xi1  = xi + 2;
      xr2  = xr + 3;
      xi2  = xi + 3;
      tmp1 = *xr1 + *xi2;
      tmp2 = *xi1 + *xr2;
      *xi1 = *xi1 - *xr2;
      *xr2 = *xr1 - *xi2;
      *xr1 = tmp1;
      *xi2 = tmp2;
      return;
   }
```

```
    else if (logm == 1) {    /* length m = 2 */
        xr2 = xr + 1;
        xi2 = xi + 1;
        tmp1 = *xr + *xr2;
        *xr2 = *xr - *xr2;
        *xr = tmp1;
        tmp1 = *xi + *xi2;
        *xi2 = *xi - *xi2;
        *xi = tmp1;
        return;
    }
    else if (logm == 0) return;    /* length m = 1 */
}

/* Compute a few constants */
m = 1 << logm; m2 = m / 2; m4 = m2 / 2; m8 = m4 /2;

/* Build tables of butterfly coefficients, if necessary */
if ((logm >= 4) && (tab[logm-4] == NULL)) {

    /* Allocate memory for tables */
    nel = m4 - 2;
    if ((tab[logm-4] = (float *) calloc(6 * nel, sizeof(float)))
        == NULL) {
        error_exit();
    }

    /* Initialize pointers */
    cn  = tab[logm-4]; spcn  = cn + nel;  smcn  = spcn + nel;
    c3n = smcn + nel; spc3n = c3n + nel; smc3n = spc3n + nel;

    /* Compute tables */
    for (n = 1; n < m4; n++) {
        if (n == m8) continue;
        ang = n * TWOPI / m;
        c = cos(ang); s = sin(ang);
        *cn++ = c; *spcn++ = - (s + c); *smcn++ = s - c;
        ang = 3 * n * TWOPI / m;
        c = cos(ang); s = sin(ang);
        *c3n++ = c; *spc3n++ = - (s + c); *smc3n++ = s - c;
    }
}

/* Step 1 */
xr1 = xr; xr2 = xr1 + m2;
xi1 = xi; xi2 = xi1 + m2;
```

```
for (n = 0; n < m2; n++) {
   tmp1 = *xr1 + *xr2;
   *xr2 = *xr1 - *xr2;
   *xr1 = tmp1;
   tmp2 = *xi1 + *xi2;
   *xi2 = *xi1 - *xi2;
   *xi1 = tmp2;
   xr1++; xr2++; xi1++; xi2++;
}

/* Step 2 */
xr1 = xr + m2; xr2 = xr1 + m4;
xi1 = xi + m2; xi2 = xi1 + m4;
for (n = 0; n < m4; n++) {
   tmp1 = *xr1 + *xi2;
   tmp2 = *xi1 + *xr2;
   *xi1 = *xi1 - *xr2;
   *xr2 = *xr1 - *xi2;
   *xr1 = tmp1;
   *xi2 = tmp2;
   xr1++; xr2++; xi1++; xi2++;
}

/* Steps 3 & 4 */
xr1 = xr + m2; xr2 = xr1 + m4;
xi1 = xi + m2; xi2 = xi1 + m4;
if (logm >= 4) {
   nel = m4 - 2;
   cn  = tab[logm-4]; spcn  = cn + nel;  smcn  = spcn + nel;
   c3n = smcn + nel;  spc3n = c3n + nel; smc3n = spc3n + nel;
}
xr1++; xr2++; xi1++; xi2++;
for (n = 1; n < m4; n++) {
   if (n == m8) {
      tmp1 =  SQHALF * (*xr1 + *xi1);
      *xi1 =  SQHALF * (*xi1 - *xr1);
      *xr1 =  tmp1;
      tmp2 =  SQHALF * (*xi2 - *xr2);
      *xi2 = -SQHALF * (*xr2 + *xi2);
      *xr2 =  tmp2;
   } else {
      tmp2 = *cn++ * (*xr1 + *xi1);
      tmp1 = *spcn++ * *xr1 + tmp2;
      *xr1 = *smcn++ * *xi1 + tmp2;
      *xi1 = tmp1;
      tmp2 = *c3n++ * (*xr2 + *xi2);
      tmp1 = *spc3n++ * *xr2 + tmp2;
      *xr2 = *smc3n++ * *xi2 + tmp2;
      *xi2 = tmp1;
   }
```

```
      xr1++; xr2++; xi1++; xi2++;
   }

   /* Call ssrec again with half DFT length */
   srrec(xr, xi, logm-1);

   /* Call ssrec again twice with one quarter DFT length.
      Constants have to be recomputed, because they are static! */
   m = 1 << logm; m2 = m / 2;
   srrec(xr + m2, xi + m2, logm-2);
   m = 1 << logm; m4 = 3 * (m / 4);
   srrec(xr + m4, xi + m4, logm-2);
}
/*-----------------------------------------------------------------*
 *      Direct transform                                           *
 *-----------------------------------------------------------------*/

void  srfft(float *xr, float *xi, int logm)
{
   /* Call recursive routine */
   srrec(xr, xi, logm);

   /* Output array unshuffling using bit-reversed indices */
   if (logm > 1) {
      BR_permute(xr, logm);
      BR_permute(xi, logm);
   }
}

/*-----------------------------------------------------------------*
 *      Inverse transform.  Uses Duhamel's trick (Ref: P. Duhamel  *
 *      et. al., "On computing the inverse DFT", IEEE Trans. ASSP, *
 *      Feb. 1988, pp. 285-286).                                   *
 *-----------------------------------------------------------------*/

void  srifft(float *xr, float *xi, int logm)
{
   int      i, m;
   float    fac, *xrp, *xip;

   /* Call direct FFT, swapping real & imaginary addresses */
   srfft(xi, xr, logm);

   /* Normalization */
   m = 1 << logm;
   fac = 1.0 / m;
   xrp = xr; xip = xi;
```

```
   for (i = 0; i < m; i++) {
      *xrp++ *= fac;
      *xip++ *= fac;
   }
}


/*-------------------------------------------------------------------*
 *    RSFFT.C   -  Split-Radix Fast Fourier Transform                *
 *                 for real-valued inputs.                           *
 *                                                                   *
 *    This is a decimation-in-frequency version, using the steps     *
 *    of the algorithm as described in Section 2.5.                  *
 *                                                                   *
 *    Author:  Henrique S. Malvar.                                   *
 *    Date:    October 8, 1991.                                      *
 *                                                                   *
 *    Usage:   rsfft(x, logm);   --  for direct DFT                  *
 *             rsifft(x, logm);  --  for inverse DFT                 *
 *                                                                   *
 *    Arguments: x (float)  - input and output vector, length M;     *
 *                            in the time domain, it contains the    *
 *                            M-point data vector; in the frequency  *
 *                            domain, it contains the real parts of  *
 *                            X(0) to X(M/2) followed by the         *
 *                            imaginary parts of X(M/2+1) to         *
 *                            X(M-1).                                 *
 *               logm (int) - log (base 2) of vector length M,       *
 *                            e.g., for M = 256 -> logm = 8.         *
 *-------------------------------------------------------------------*/

#include <stdio.h>
#include <stdlib.h>
#include <alloc.h>
#include <math.h>

#define   MAXLOGM    11    /* max FFT length = 2^MAXLOGM */
#define   TWOPI      6.28318530717958647692
#define   SQHALF     0.707106781186547524401

/*-------------------------------------------------------------------*
 *    Prototypes of external functions.                              *
 *-------------------------------------------------------------------*/

void BR_permute(float *x, int logm);
void srrec(float *xr, float *xi, int logm);

/*-------------------------------------------------------------------*
 *    Error exit for program abortion.                               *
 *-------------------------------------------------------------------*/
```

```
static   void   error_exit(void)
{
   exit(1);
}
/*----------------------------------------------------------------*
 *    Recursive part of the RSFFT algorithm.  Not externally       *
 *    callable.                                                     *
 *----------------------------------------------------------------*/
static void rsrec(float *x, int logm)
{
   static   int       m, m2, m4, m8, nel, n;
   static   float     *xr1, *xr2, *xi1;
   static   float     *cn, *spcn, *smcn;
   static   float     tmp1, tmp2, ang, c, s;
   static   float     *tab[MAXLOGM];

   /* Check range of logm */
   if ((logm < 0) || (logm > MAXLOGM)) {
      printf("Error : RSFFT : logm = %d is out of bounds [%d, %d]\n",
         logm, 0, MAXLOGM);
      error_exit();
   }

   /* Compute trivial cases */
   if (logm < 2) {
      if (logm == 1) {   /* length m = 2 */
         xr2  = x + 1;
         tmp1 = *x + *xr2;
         *xr2 = *x - *xr2;
         *x   = tmp1;
         return;
      }
      else if (logm == 0) return;   /* length m = 1 */
   }

   /* Compute a few constants */
   m = 1 << logm; m2 = m / 2; m4 = m2 / 2; m8 = m4 /2;

   /* Build tables of butterfly coefficients, if necessary */
   if ((logm >= 4) && (tab[logm-4] == NULL)) {

      /* Allocate memory for tables */
      nel = m4 - 2;
```

```
    if ((tab[logm-4] = (float *) calloc(3 * nel, sizeof(float)))
        == NULL) {
        printf("Error : RSFFT : not enough memory for cosine tables.\n");
        error_exit();
    }

    /* Initialize pointers */
    cn  = tab[logm-4]; spcn  = cn + nel;  smcn = spcn + nel;

    /* Compute tables */
    for (n = 1; n < m4; n++) {
        if (n == m8) continue;
        ang = n * TWOPI / m;
        c = cos(ang); s = sin(ang);
        *cn++ = c; *spcn++ = - (s + c); *smcn++ = s - c;
    }
}

/* Step 1 */
xr1 = x; xr2 = xr1 + m2;
for (n = 0; n < m2; n++) {
    tmp1 = *xr1 + *xr2;
    *xr2 = *xr1 - *xr2;
    *xr1 = tmp1;
    xr1++; xr2++;
}

/* Step 2 */
xr1 = x + m2 + m4;
for (n = 0; n < m4; n++) {
    *xr1 = - *xr1;
    xr1++;
}

/* Steps 3 & 4 */
xr1 = x + m2; xi1 = xr1 + m4;
if (logm >= 4) {
    nel = m4 - 2;
    cn  = tab[logm-4]; spcn  = cn + nel;  smcn = spcn + nel;
}
xr1++; xi1++;
for (n = 1; n < m4; n++) {
    if (n == m8) {
        tmp1 =  SQHALF * (*xr1 + *xi1);
        *xi1 =  SQHALF * (*xi1 - *xr1);
        *xr1 =  tmp1;
    } else {
        tmp2 = *cn++ * (*xr1 + *xi1);
```

```
            tmp1 = *spcn++ * *xr1 + tmp2;
            *xr1 = *smcn++ * *xi1 + tmp2;
            *xi1 = tmp1;
        }
        xr1++; xi1++;
    }

    /* Call rsrec again with half DFT length */
    rsrec(x, logm-1);

    /* Call complex DFT routine, with quarter DFT length.
       Constants have to be recomputed, because they are static! */
    m = 1 << logm; m2 = m / 2; m4 = 3 * (m / 4);
    srrec(x + m2, x + m4, logm-2);

    /* Step 5: sign change & data reordering */
    m = 1 << logm; m2 = m / 2; m4 = m2 / 2; m8 = m4 / 2;
    xr1 = x + m2 + m4;
    xr2 = x + m - 1;
    for (n = 0; n < m8; n++) {
        tmp1    =    *xr1;
        *xr1++ =  - *xr2;
        *xr2-- =  - tmp1;
    }
    xr1 = x + m2 + 1;
    xr2 = x + m - 2;
    for (n = 0; n < m8; n++) {
        tmp1    =    *xr1;
        *xr1++ =  - *xr2;
        *xr2-- =    tmp1;
        xr1++;
        xr2--;
    }
    if (logm == 2) x[3] = -x[3];
}

/*------------------------------------------------------------------------*
 *      Direct transform for real inputs                                  *
 *------------------------------------------------------------------------*/

void  rsfft(float *x, int logm)
{
    /* Call recursive routine */
    rsrec(x, logm);

    /* Output array unshuffling using bit-reversed indices */
    if (logm > 1) {
        BR_permute(x, logm);
    }
}
```

```
/*------------------------------------------------------------------*
 *     Recursive part of the inverse RSFFT algorithm.  Not externally *
 *     callable.                                                      *
 *------------------------------------------------------------------*/
static void rsirec(float *x, int logm)
{
   static   int     m, m2, m4, m8, nel, n;
   static   float   *xr1, *xr2, *xi1;
   static   float   *cn, *spcn, *smcn;
   static   float   tmp1, tmp2, ang, c, s;
   static   float   *tab[MAXLOGM];

   /* Check range of logm */
   if ((logm < 0) || (logm > MAXLOGM)) {
      printf("Error : RSFFT : logm = %d is out of bounds [%d, %d]\n",
         logm, 0, MAXLOGM);
      error_exit();
   }

   /* Compute trivial cases */
   if (logm < 2) {
      if (logm == 1) {   /* length m = 2 */
         xr2  = x + 1;
         tmp1 = *x + *xr2;
         *xr2 = *x - *xr2;
         *x   = tmp1;
         return;
      }
      else if (logm == 0) return;   /* length m = 1 */
   }

   /* Compute a few constants */
   m = 1 << logm; m2 = m / 2; m4 = m2 / 2; m8 = m4 /2;

   /* Build tables of butterfly coefficients, if necessary */
   if ((logm >= 4) && (tab[logm-4] == NULL)) {

      /* Allocate memory for tables */
      nel = m4 - 2;
      if ((tab[logm-4] = (float *) calloc(3 * nel, sizeof(float)))
         == NULL) {
         printf("Error : RSFFT : not enough memory for cosine tables.\n");
         error_exit();
      }

      /* Initialize pointers */
      cn  = tab[logm-4]; spcn  = cn + nel;  smcn  = spcn + nel;
```

```
    /* Compute tables */
    for (n = 1; n < m4; n++) {
        if (n == m8) continue;
        ang = n * TWOPI / m;
        c = cos(ang); s = sin(ang);
        *cn++ = c; *spcn++ = - (s + c); *smcn++ = s - c;
    }
}

/* Reverse Step 5: sign change & data reordering */
if (logm == 2) x[3] = -x[3];
xr1 = x + m2 + 1;
xr2 = x + m - 2;
for (n = 0; n < m8; n++) {
    tmp1     =     *xr1;
    *xr1++ =     *xr2;
    *xr2-- = - tmp1;
    xr1++;
    xr2--;
}
xr1 = x + m2 + m4;
xr2 = x + m - 1;
for (n = 0; n < m8; n++) {
    tmp1     =     *xr1;
    *xr1++ = - *xr2;
    *xr2-- = - tmp1;
}

/* Call rsirec again with half DFT length */
rsirec(x, logm-1);

/* Call complex DFT routine, with quarter DFT length.
   Constants have to be recomputed, because they are static! */
m = 1 << logm; m2 = m / 2; m4 = 3 * (m / 4);
srrec(x + m4, x + m2, logm-2);

/* Reverse Steps 3 & 4 */
m = 1 << logm; m2 = m / 2; m4 = m2 / 2; m8 = m4 /2;
xr1 = x + m2; xi1 = xr1 + m4;
if (logm >= 4) {
    nel = m4 - 2;
    cn  = tab[logm-4]; spcn  = cn + nel;  smcn  = spcn + nel;
}
xr1++; xi1++;
for (n = 1; n < m4; n++) {
    if (n == m8) {
        tmp1 = SQHALF * (*xr1 - *xi1);
        *xi1 = SQHALF * (*xi1 + *xr1);
        *xr1 =  tmp1;
```

```
    } else {
        tmp2 = *cn++ * (*xr1 + *xi1);
        tmp1 = *smcn++ * *xr1 + tmp2;
        *xr1 = *spcn++ * *xi1 + tmp2;
        *xi1 = tmp1;
    }
    xr1++; xi1++;
}

/* Reverse Step 2 */
xr1 = x + m2 + m4;
for (n = 0; n < m4; n++) {
    *xr1 = - *xr1;
    xr1++;
}

/* Reverse Step 1 */
xr1 = x; xr2 = xr1 + m2;
for (n = 0; n < m2; n++) {
    tmp1 = *xr1 + *xr2;
    *xr2 = *xr1 - *xr2;
    *xr1 = tmp1;
    xr1++; xr2++;
}
}

/*------------------------------------------------------------------------*
 *      Inverse transform for real inputs                                 *
 *------------------------------------------------------------------------*/

void  rsifft(float *x, int logm)
{
    int      i, m;
    float    fac, *xp;

    /* Output array unshuffling using bit-reversed indices */
    if (logm > 1) {
        BR_permute(x, logm);
    }
    x[0] *= 0.5;
    if (logm > 0) x[1] *= 0.5;

    /* Call recursive routine */
    rsirec(x, logm);

    /* Normalization */
    m = 1 << logm;
    fac = 2.0 / m;
    xp = x;
```

```c
   for (i = 0; i < m; i++) {
      *xp++ *= fac;
   }
}

/*----------------------------------------------------------------*
 *    FDHT.C  -  Split-Radix Fast Hartley Transform.              *
 *                                                                *
 *    This is a decimation-in-frequency version, using the steps  *
 *    of the algorithm as described in Section 2.5.               *
 *                                                                *
 *    Author:  Henrique S. Malvar.                                *
 *    Date:    October 8, 1991.                                   *
 *                                                                *
 *    Usage:   fdht(x, logm);   -- for direct or inverse DHT      *
 *                                                                *
 *    Arguments: x (float) - input and output vector, length M;   *
 *               logm (int) - log (base 2) of vector length M,    *
 *                            e.g., for M = 256 -> logm = 8.      *
 *----------------------------------------------------------------*/

#include <stdio.h>
#include <stdlib.h>
#include <alloc.h>
#include <math.h>

#define   MAXLOGM   11      /* max DHT length = 2^MAXLOGM */
#define   TWOPI     6.28318530717958647692
#define   SQHALF    0.707106781186547524401
#define   SQTWO     1.414213562373095048 80

/*----------------------------------------------------------------*
 *    Error exit for program abortion.                            *
 *----------------------------------------------------------------*/

static   void   error_exit(void)
{
   exit(1);
}

/*----------------------------------------------------------------*
 *    Data unshuffling according to bit-reversed indexing.        *
 *                                                                *
 *    Bit reversal is done using Evan's algorithm (Ref: D. M. W.  *
 *    Evans, "An improved digit-reversal permutation algorithm ...", *
 *    IEEE Trans. ASSP, Aug. 1987, pp. 1120-1125).                *
 *----------------------------------------------------------------*/

static   int   brseed[256];   /* Evans' seed table */
static   int   brsflg;        /* flag for table building */
```

```
static void BR_permute(float *x, int logm)
{
    int     i, j, imax, lg2, n;
    int     off, fj, gno, *brp;
    float   tmp, *xp, *xq;

    lg2 = logm >> 1;
    n = 1 << lg2;
    if (logm & 1) lg2++;

    /* Create seed table if not yet built */
    if (brsflg != logm) {
        brsflg = logm;
        brseed[0] = 0;
        brseed[1] = 1;
        for (j = 2; j <= lg2; j++) {
            imax = 1 << (j - 1);
            for (i = 0; i < imax; i++) {
                brseed[i] <<= 1;
                brseed[i + imax] = brseed[i] + 1;
            }
        }
    }

    /* Unshuffling loop */
    for (off = 1; off < n; off++) {
        fj = n * brseed[off]; i = off; j = fj;
        tmp = x[i]; x[i] = x[j]; x[j] = tmp;
        xp = &x[i];
        brp = &brseed[1];
        for (gno = 1; gno < brseed[off]; gno++) {
            xp += n;
            j = fj + *brp++;
            xq = x + j;
            tmp = *xp; *xp = *xq; *xq = tmp;
        }
    }
}

/*----------------------------------------------------------------------*
 *      Recursive part of the FHT algorithm.  Not externally            *
 *      callable.                                                       *
 *----------------------------------------------------------------------*/

static void dhtrec(float *x, int logm)
{
    static  int     m, m2, m4, m8, nel, n;
    static  float   *x1, *x2, *x3, *x4;
    static  float   *sn, *cpsn, *cmsn, *s3n, *cps3n, *cms3n;
```

```
static    float     tmp1, tmp2, tmp3, ang, c, s;
static    float     *tab[MAXLOGM];

/* Check range of logm */
if ((logm < 0) || (logm > MAXLOGM)) {
   printf("Error : FDHT : logm = %d is out of bounds [%d, %d]\n",
      logm, 0, MAXLOGM);
   error_exit();
}

/* Compute trivial cases */
if (logm <= 2) {
   if (logm == 2) {  /* length m = 4 */
      x1 = x;   x2 = x + 2;
      tmp1 = *x1 + *x2;
      *x2  = *x1 - *x2;
      *x1  = tmp1;
      x1++;   x2++;
      tmp1 = *x1 + *x2;
      *x2  = *x1 - *x2;
      *x1  = tmp1;
      x1 = x;   x2 = x + 1;
      tmp1 = *x1 + *x2;
      *x2  = *x1 - *x2;
      *x1  = tmp1;
      x1++;   x2++;
      x1++;   x2++;
      tmp1 = *x1 + *x2;
      *x2  = *x1 - *x2;
      *x1  = tmp1;
      return;
   } else if (logm == 1) {  /* length m = 2 */
      x2  = x + 1;
      tmp1 = *x + *x2;
      *x2  = *x - *x2;
      *x   = tmp1;
      return;
   }
   else if (logm == 0) return;   /* length m = 1 */
}

/* Compute a few constants */
m = 1 << logm; m2 = m / 2; m4 = m2 / 2; m8 = m4 /2;

/* Build tables of butterfly coefficients, if necessary */
if ((logm >= 4) && (tab[logm-4] == NULL)) {

   /* Allocate memory for tables */
   nel = m8 - 1;
```

```
    if ((tab[logm-4] = (float *) calloc(6 * nel, sizeof(float)))
        == NULL) {
        printf("Error : FDHT : not enough memory for cosine tables.\n");
        error_exit();
    }

    /* Initialize pointers */
    sn  = tab[logm-4]; cpsn  = sn + nel;   cmsn  = cpsn + nel;
    s3n = cmsn + nel;  cps3n = s3n + nel; cms3n = cps3n + nel;

    /* Compute tables */
    for (n = 1; n < m8; n++) {
        ang = n * TWOPI / m;
        c = cos(ang); s = sin(ang);
        *sn++ = s; *cpsn++ = - (c + s); *cmsn++ = c - s;
        ang = 3 * n * TWOPI / m;
        c = cos(ang); s = sin(ang);
        *s3n++ = s; *cps3n++ = - (c + s); *cms3n++ = c - s;
    }
}

/* Step 1 */
x1 = x; x2 = x1 + m2;
for (n = 0; n < m2; n++) {
    tmp1 = *x1 + *x2;
    *x2  = *x1 - *x2;
    *x1  = tmp1;
    x1++; x2++;
}

/* Step 2 */
x1 = x + m2; x2 = x1 + m4;
for (n = 0; n < m8; n++) {
    tmp1 = *x1 + *x2;
    *x2  = *x1 - *x2;
    *x1  = tmp1;
    x1++; x2--;
}

/* Step 3 */
x1 = x + m2 + m4 + 1; x2 = x + m - 1;
for (n = 1; n < m8; n++) {
    tmp1 = *x2 + *x1;
    *x2  = *x2 - *x1;
    *x1  = tmp1;
    x1++; x2--;
}
```

```
/* Step 4 */
x[m2 + m8]      *= SQTWO;
x[m2 + m4 + m8] *= SQTWO;

/* Step 5 */
if (logm >= 4) {
   nel = m8 - 1;
   sn  = tab[logm-4]; cpsn  = sn + nel;   cmsn  = cpsn + nel;
   s3n = cmsn + nel;  cps3n = s3n + nel;  cms3n = cps3n + nel;
}
x1 = x + m2;  x2 = x1 + m4;  x3 = x2;  x4 = x + m;
x1++; x2--; x3++; x4--;
for (n = 1; n < m8; n++) {
   tmp2 = *sn++ * (*x1 + *x4);
   *x1  = *cmsn++ * *x1 + tmp2;
   tmp3 = *cpsn++ * *x4 + tmp2;
   tmp2 = *s3n++ * (*x2 + *x3);
   *x4  = *cps3n++ * *x3 + tmp2;
   *x3  = *cms3n++ * *x2 + tmp2;
   *x2  = tmp3;
   x1++; x2--; x3++; x4--;
}

/* Call dhtrec again with half length */
dhtrec(x, logm-1);

/* Call dhtrec again twice with one quarter length.
   Constants have to be recomputed, because they are static! */
m = 1 << logm; m2 = m / 2;
dhtrec(x + m2, logm-2);
m = 1 << logm; m4 = 3 * (m / 4);
dhtrec(x + m4, logm-2);
}
/*-----------------------------------------------------------------------*
 *     Externally callable module.                                       *
 *-----------------------------------------------------------------------*/

void  fdht(float *x, int logm)
{
   int    i, m;
   float  fac, *xp;

   /* Call recursive routine */
   dhtrec(x, logm);

   /* Output array unshuffling using bit-reversed indices */
   if (logm > 1) {
      BR_permute(x, logm);
   }
```

```
    /* Normalization */
    m = 1 << logm;
    fac = sqrt(1.0 / m);
    xp = x;
    for (i = 0; i < m; i++) {
       *xp++ *= fac;
    }
}

/*------------------------------------------------------------------------*
 *    FDCTIV.C  -  Fast Discrete Cosine Transform, type IV                *
 *                                                                        *
 *    The computation is done via a half-length FFT, according           *
 *    to the algorithm described in Section 2.5.                         *
 *                                                                        *
 *    Author:  Henrique S. Malvar.                                       *
 *    Date:    October 8, 1991.                                          *
 *                                                                        *
 *    Usage:   fdctiv(x, logm);   -- direct or inverse transform          *
 *                                                                        *
 *    Arguments:  x (float)   - input and output vector, length M.       *
 *                logm (int)  - log (base 2) of vector length M,          *
 *                              e.g., for M = 256 -> logm = 8.            *
 *                                                                        *
 *    Note: this is an orthogonal transform that is also symmetric.      *
 *    Thus, it is its own inverse.                                       *
 *------------------------------------------------------------------------*/

#include <stdio.h>
#include <stdlib.h>
#include <alloc.h>
#include <math.h>

#define   MAXLOGM     12      /* max DCT-IV length = 2^MAXLOGM */
#define   TWOPI       6.28318530717958647692
#define   PI          3.14159265358979323846
#define   SQHALF      0.707106781186547524401
#define   COSM2       0.92387953251129
#define   SINM2       0.38268343236509

void  srfft(float *, float *, int); /* srfft prototype */

/*------------------------------------------------------------------------*
 *    Error exit for program abortion.                                   *
 *------------------------------------------------------------------------*/

static   void  error_exit(void)
{
    exit(1);
}
```

```
/*----------------------------------------------------------------------*
 *     DCT-IV transform                                                  *
 *----------------------------------------------------------------------*/

void  fdctiv(float *x, int logm)
{
    int        i, m, m2, m4, n, nel;
    static     float    *y, *xp1, *xp2, *yp1, *yp2;
    static     float    *c4n, *spc4n, *smc4n, *cn, *spcn, *smcn;
    static     float    tmp1, tmp2, ang, c, s, fac;
    static     float    *tab[MAXLOGM];
    static     float    *yt[MAXLOGM];

    /* Check range of logm */
    if ((logm < 0) || (logm > MAXLOGM)) {
        printf("Error : FDCTIV : logm = %d is out of bounds [%d, %d]\n",
            logm, 0, MAXLOGM);
        error_exit();
    }

    /* Trivial cases, m = 1 and m = 2 */
    if (logm < 1) return;

    if (logm == 1) {
        xp2 = x + 1;
        tmp1 = COSM2 * *x + SINM2 * *xp2;
        *xp2 = SINM2 * *x - COSM2 * *xp2;
        *x   = tmp1;
        return;
    }

    /* Compute m */
    m = 1 << logm;
    m2 = m / 2;
    m4 = m2 / 2;

    /* Build tables of butterfly coefficients, if necessary */
    if (tab[logm-2] == NULL) {

        /* Allocate memory for tables */
        nel = m2;
        if ((tab[logm-2] = (float *) calloc(6 * nel, sizeof(float)))
            == NULL) {
            printf("Error : FDCTIV : not enough memory for cosine tables.\n");
            error_exit();
        }

        /* Initialize pointers */
        c4n = tab[logm-2]; spc4n = c4n + nel; smc4n = spc4n + nel;
        cn  = smc4n + nel; spcn  = cn + nel;  smcn  = spcn + nel;
```

```
/* Compute tables. For the tables in the second modulation
   stage, all entries should be multiplied by sqrt(2/m), in
   order to generate an orthogonal transform */
fac = sqrt(2.0 / m);
for (n = 0; n < m2; n++) {
    ang = (n + 0.25) * PI / m;
    c = cos(ang); s = sin(ang);
    *c4n++ = c; *spc4n++ = - (s + c); *smc4n++ = s - c;
    if (n == m4) {
        c = fac * SQHALF;
        *cn++ = c;
    } else {
        ang = n * PI / m;
        c = fac * cos(ang); s = fac * sin(ang);
        *cn++ = c; *spcn++ = - (s + c); *smcn++ = s - c;
    }
}
}

/* Allocate space for working vector, if necessary */
if (yt[logm-2] == NULL) {
    if ((yt[logm-2] = (float *) calloc(m2, sizeof(float))) == NULL) {
        printf("Error : FDCTIV: not enough memory for working vector.\n");
        error_exit();
    }
}

/* Define table pointers */
nel = m2;
c4n = tab[logm-2]; spc4n = c4n + nel; smc4n = spc4n + nel;
cn  = smc4n + nel; spcn  = cn + nel;   smcn  = spcn + nel;
y   = yt[logm-2];

/* Step 1: input data reordering */
xp1 = x; xp2 = x + 1;
yp1 = x; yp2 = y + m2 - 1;
for (i = 0; i < m2; i++) {
    *yp2-- = *xp2++;
    *yp1++ = *xp1++;
    xp1++; xp2++;
}

/* Step 2: first modulation stage */
yp1 = x; yp2 = y;
for (i = 0; i < m2; i++) {
    tmp2 = *c4n++ * (*yp1 + *yp2);
    tmp1 = *spc4n++ * *yp1 + tmp2;
    *yp1 = *smc4n++ * *yp2 + tmp2;
    *yp2 = tmp1;
```

```
        yp1++; yp2++;
    }

    /* Step 3: call FFT of half length */
    srfft(x, y, logm-1);

    /* Step 4: second modulation stage */
    yp1 = x; yp2 = y;
    for (i = 0; i < m2; i++) {
        if (i == m4 ) {
            tmp1 =  *cn * (*yp1 + *yp2);
            *yp2 =  *cn++ * (*yp2 - *yp1);
            *yp1 =  tmp1;
        } else {
            tmp2 = *cn++ * (*yp1 + *yp2);
            tmp1 = *spcn++ * *yp1 + tmp2;
            *yp1 = *smcn++ * *yp2 + tmp2;
            *yp2 = tmp1;
        }
        yp1++; yp2++;
    }

    /* Step 5: output data reordering */
    xp1 = x; xp2 = &x[m-1];
    yp1 = y; yp2 = y + m2;

    xp1 = x + m - 2;  xp2 = x + m - 1;
    yp1 = x + m2 - 1; yp2 = y;
    for (i = 0; i < m2; i++) {
        *xp1-- =   *yp1--;
        *xp2-- = - *yp2++;
        xp1--; xp2--;
    }
}
```

```
/*-------------------------------------------------------------------------*
 *     FDCTIV2.C  -  Fast Discrete Cosine Transform, type IV               *
 *                                                                         *
 *     Version 2: with scaling factor.                                     *
 *                                                                         *
 *     The computation is done via a half-length FFT, according            *
 *     to the algorithm described in Section 2.5.                          *
 *                                                                         *
 *     Author:  Henrique S. Malvar.                                        *
 *     Date:    October 8, 1991.                                           *
 *                                                                         *
 *     Usage:   fdctiv2(x, logm, fac);  -- direct or inverse transform *
 *                                                                         *
 *     Arguments:  x (float)    - input and output vector, length M.       *
 *                 logm (int)   - log (base 2) of vector length M,         *
 *                                e.g., for M = 256 -> logm = 8.           *
 *                 fac (float)  - scaling factor. If fac == 1, an          *
 *                                orthogonal transform is performed;       *
 *                                otherwise, outputs are multiplied        *
 *                                by fac; see note below.                  *
 *                                                                         *
 *     Note: the scaling factor argument fac is only used the first        *
 *     time the routine is called with a new value for logm.               *
 *     Subsequent calls will use the previously defined value of fac.      *
 *                                                                         *
 *     This version of the fdctiv is called by the fdct module.            *
 *-------------------------------------------------------------------------*/

#include <stdio.h>
#include <stdlib.h>
#include <alloc.h>
#include <math.h>

#define   MAXLOGM     12    /* max DCT-IV length = 2^MAXLOGM */
#define   TWOPI       6.28318530717958647692
#define   PI          3.14159265358979323846
#define   SQHALF      0.707106781186547524401
#define   COSM2       0.92387953251129
#define   SINM2       0.38268343236509

void  srfft(float *, float *, int); /* srfft prototype */

/*-------------------------------------------------------------------------*
 *     Error exit for program abortion.                                    *
 *-------------------------------------------------------------------------*/

static   void  error_exit(void)
{
   exit(1);
}
```

```
/*----------------------------------------------------------------------*
 *    DCT-IV transform                                                   *
 *----------------------------------------------------------------------*/

void   fdctiv2(float *x, int logm, float fac)
{
   int       i, m, m2, m4, n, nel;
   static    float    *y, *xp1, *xp2, *yp1, *yp2;
   static    float    *c4n, *spc4n, *smc4n, *cn, *spcn, *smcn;
   static    float    tmp1, tmp2, ang, c, s;
   static    float    *tab[MAXLOGM];
   static    float    *yt[MAXLOGM];

   /* Check range of logm */
   if ((logm < 0) || (logm > MAXLOGM)) {
      printf("Error : FDCTIV : logm = %d is out of bounds [%d, %d]\n",
         logm, 0, MAXLOGM);
      error_exit();
   }

   /* Trivial cases, m = 1 and m = 2 */
   if (logm < 1) {
      *x *= fac;
      return;
   }

   if (logm == 1) {
      if (tab[0] == NULL) {
         if ((tab[0] = (float *) calloc(2, sizeof(float))) == NULL) {
            printf("Error : FDCTIV2 : not enough memory for cosine"
               " tables.\n");
            error_exit();
         }
         yp1 = tab[0];
         *yp1++ = fac * COSM2;
         *yp1    = fac * SINM2;
      }
      xp2 = x + 1;
      yp1 = tab[0]; yp2 = yp1 + 1;
      tmp1 = *yp1 * *x + *yp2 * *xp2;
      *xp2 = *yp2 * *x - *yp1 * *xp2;
      *x    = tmp1;
      return;
   }

   /* Compute m */
   m = 1 << logm;
   m2 = m / 2;
   m4 = m2 / 2;
```

```
/* Build tables of butterfly coefficients, if necessary */
if (tab[logm-1] == NULL) {

    /* Allocate memory for tables */
    nel = m2;
    if ((tab[logm-1] = (float *) calloc(6 * nel, sizeof(float)))
        == NULL) {
        printf("Error : FDCTIV2: not enough memory for"
            " cosine tables.\n");
        error_exit();
    }

    /* Initialize pointers */
    c4n = tab[logm-1]; spc4n = c4n + nel; smc4n = spc4n + nel;
    cn  = smc4n + nel;  spcn  = cn + nel;  smcn  = spcn + nel;

    /* Compute tables. For the tables in the second modulation
       stage, all entries should be multiplied by sqrt(2/m), in
       order to generate an orthogonal transform */
    fac *= sqrt(2.0 / m);
    for (n = 0; n < m2; n++) {
        ang = (n + 0.25) * PI / m;
        c = cos(ang); s = sin(ang);
        *c4n++ = c; *spc4n++ = - (s + c); *smc4n++ = s - c;
        if (n == m4) {
            c = fac * SQHALF;  *cn++ = c;
        } else {
            ang = n * PI / m;
            c = fac * cos(ang); s = fac * sin(ang);
            *cn++ = c; *spcn++ = - (s + c); *smcn++ = s - c;
        }
    }
}

/* Allocate space for working vector, if necessary */
if (yt[logm-1] == NULL) {
    if ((yt[logm-1] = (float *) calloc(m2, sizeof(float))) == NULL) {
        printf("Error : FDCTIV2: not enough memory for"
            " working vector.\n");
        error_exit();
    }
}

/* Define table pointers */
nel = m2;
c4n = tab[logm-1]; spc4n = c4n + nel; smc4n = spc4n + nel;
cn  = smc4n + nel; spcn  = cn + nel;  smcn  = spcn + nel;
y   = yt[logm-1];
```

```
/* Step 1: input data reordering */
xp1 = x; xp2 = x + 1;
yp1 = x; yp2 = y + m2 - 1;
for (i = 0; i < m2; i++) {
   *yp2-- = *xp2++;
   *yp1++ = *xp1++;
   xp1++; xp2++;
}

/* Step 2: first modulation stage */
yp1 = x; yp2 = y;
for (i = 0; i < m2; i++) {
   tmp2 = *c4n++ * (*yp1 + *yp2);
   tmp1 = *spc4n++ * *yp1 + tmp2;
   *yp1 = *smc4n++ * *yp2 + tmp2;
   *yp2 = tmp1;
   yp1++; yp2++;
}

/* Step 3: call FFT of half length */
srfft(x, y, logm-1);

/* Step 4: second modulation stage */
yp1 = x; yp2 = y;
for (i = 0; i < m2; i++) {
   if (i == m4 ) {
      tmp1 = *cn * (*yp1 + *yp2);
      *yp2 = *cn++ * (*yp2 - *yp1);
      *yp1 = tmp1;
   } else {
      tmp2 = *cn++ * (*yp1 + *yp2);
      tmp1 = *spcn++ * *yp1 + tmp2;
      *yp1 = *smcn++ * *yp2 + tmp2;
      *yp2 = tmp1;
   }
   yp1++; yp2++;
}

/* Step 5: output data reordering */
xp1 = x; xp2 = &x[m-1];
yp1 = y; yp2 = y + m2;

xp1 = x + m - 2;  xp2 = x + m - 1;
yp1 = x + m2 - 1; yp2 = y;
for (i = 0; i < m2; i++) {
   *xp1-- =   *yp1--;
   *xp2-- = - *yp2++;
   xp1--; xp2--;
}
}
```

```
/*-----------------------------------------------------------------------*
 *      FDCT.C  -  Fast Discrete Cosine Transform.                       *
 *                                                                       *
 *      The computation is done via a half-length DCT and a half-length  *
 *      DCT-IV, according to the algorithm described in Section 2.5.     *
 *                                                                       *
 *      In order to minimize the number of multiplications, the          *
 *      transform is not orthogonal.  All outputs are multiplied by      *
 *      sqrt(M).                                                         *
 *                                                                       *
 *      Author:  Henrique S. Malvar.                                     *
 *      Date:    October 8, 1991.                                        *
 *                                                                       *
 *      Usage:   fdct(x, logm);    -- for direct transform               *
 *               fidct(x, logm);   -- for inverse transform              *
 *                                                                       *
 *      Arguments:  x (float)  - input and output vector, length M.      *
 *                  logm (int) - log (base 2) of vector length M,        *
 *                               e.g., for M = 256 -> logm = 8.          *
 *-----------------------------------------------------------------------*/

#include <stdio.h>
#include <stdlib.h>
#include <string.h>
#include <alloc.h>
#include <math.h>

#define  MAXLOGM      12    /* max DCT length = 2^MAXLOGM */
#define  SQHALF        0.707106781186547524401
#define  SQ2           1.414213562373095048880

void  fdctiv2(float *, int, float); /* fdctiv2 prototype */

/*-----------------------------------------------------------------------*
 *      Pointers for working vectors.                                    *
 *-----------------------------------------------------------------------*/

static float *yt[MAXLOGM];

/*-----------------------------------------------------------------------*
 *      Error exit for program abortion.                                 *
 *-----------------------------------------------------------------------*/

static   void  error_exit(void)
{
   exit(1);
}
```

```
/*-----------------------------------------------------------------------*
 *      Recursive part of the DCT algorithm.  Not externally             *
 *      callable.                                                        *
 *-----------------------------------------------------------------------*/
static void dctrec(float *x, int logm)
{
    static   float    tmp, *x1, *x2;
    static   int      n, m, m2, m4;

    /* Stop recursion when m = 2 */
    if (logm == 1) {
        x2 = x + 1;
        tmp = (*x + *x2);
        *x2 = (*x - *x2);
        *x  = tmp;
        return;
    }

    m = 1 << logm;
    m2 = m / 2;
    m4 = m2 / 2;

    /* +/- butterflies (see Fig. 2.18) */
    x1 = x; x2 = x1 + m - 1;
    for (n = 0; n < m2; n++) {
        tmp = *x1 + *x2;
        *x2 = *x1 - *x2;
        *x1 = tmp;
        x1++; x2--;
    }

    /* Swap entries of bottom half vector */
    x1 = x + m2; x2 = x + m - 1;
    for (n = 0; n < m4; n++) {
        tmp = *x2;
        *x2 = *x1;
        *x1 = tmp;
        x1++; x2--;
    }

    /* DCT-IV on the second half of x */
    fdctiv2(x + m2, logm-1, sqrt(m2));

    /* DCT on the first half of x */
    dctrec(x, logm-1);
}
```

```
/*------------------------------------------------------------------*
 *    Recursive part of the IDCT algorithm.  Not externally         *
 *    callable.                                                     *
 *------------------------------------------------------------------*/
static void idctrec(float *x, int logm)
{
    static   float    tmp, *x1, *x2;
    static   int      n, m, m2, m4;

    /* Stop recursion when m = 2 */
    if (logm == 1) {
        x2 = x + 1;
        tmp = (*x + *x2);
        *x2 = (*x - *x2);
        *x  = tmp;
        return;
    }

    /* DCT-IV on the second half of x */
    m = 1 << logm;
    m2 = m / 2;
    fdctiv2(x + m2, logm-1, sqrt(m2));

    /* IDCT on the first half of x */
    idctrec(x, logm-1);

    m = 1 << logm;
    m2 = m / 2;
    m4 = m2 / 2;

    /* Swap entries of bottom half vector */
    x1 = x + m2; x2 = x + m - 1;
    for (n = 0; n < m4; n++) {
        tmp = *x2;
        *x2 = *x1;
        *x1 = tmp;
        x1++; x2--;
    }

    /* +/- butterflies (see Fig. 2.18) */
    x1 = x; x2 = x1 + m - 1;
    for (n = 0; n < m2; n++) {
        tmp = *x1 + *x2;
        *x2 = *x1 - *x2;
        *x1 = tmp;
        x1++; x2--;
    }
}
```

```
/*----------------------------------------------------------------*
 *     Direct transform                                           *
 *----------------------------------------------------------------*/

void  fdct(float *x, int logm)
{
    static   float     tmp, *xp, *y, *yp;
    static   int       gr, n, m, m2, dx;

    /* Check range of logm */
    if ((logm < 0) || (logm > MAXLOGM)) {
        printf("Error : FDCT : logm = %d is out of bounds [%d, %d]\n",
            logm, 0, MAXLOGM);
        error_exit();
    }

    /* Trivial cases, m = 1 and m = 2 */
    if (logm < 1) return;

    if (logm == 1) {
        xp  = x + 1;
        tmp = (*x + *xp);
        *xp = (*x - *xp);
        *x  = tmp;
        return;
    }

    m = 1 << logm;
    m2 = m / 2;

    /* Allocate space for working vector, if necessary */
    if (yt[logm-1] == NULL) {
        if ((yt[logm-1] = (float *) calloc(m, sizeof(float))) == NULL) {
            printf("Error : FDCT : not enough memory for working vector.\n");
            error_exit();
        }
    }
    y = yt[logm-1];

    /* Copy x into y */
    memcpy(y, x, sizeof(float) * m);

    /* Call recursive module */
    dctrec(y, logm);

    /* Copy y back into x, with data unshuffling */
    dx = 2;
```

```
    for (gr = 0; gr < logm; gr++) {
        xp = x + (1 << gr); yp = y + m2;
        for (n = 0; n < m2; n++) {
            *xp = *yp++;
            xp += dx;
        }
        m2 >>= 1;
        dx <<= 1;
    }
    x[0]   = y[0];
    x[m/2] = y[1];
}

/*----------------------------------------------------------------------*
 *      Inverse transform                                               *
 *----------------------------------------------------------------------*/

void  fidct(float *x, int logm)
{
    static    float    tmp, *xp, *y, *yp;
    static    int      gr, n, m, m2, dx;

    /* Check range of logm */
    if ((logm < 0) || (logm > MAXLOGM)) {
        printf("Error : FIDCT : logm = %d is out of bounds [%d, %d]\n",
            logm, 0, MAXLOGM);
        error_exit();
    }

    /* Trivial cases, m = 1 and m = 2 */
    if (logm < 1) return;

    if (logm == 1) {
        xp  = x + 1;
        tmp = 0.5 * (*x + *xp);
        *xp = 0.5 * (*x - *xp);
        *x  = tmp;
        return;
    }

    m = 1 << logm;
    m2 = m / 2;

    /* Allocate space for working vector, if necessary */
    if (yt[logm-1] == NULL) {
        if ((yt[logm-1] = (float *) calloc(m, sizeof(float))) == NULL) {
            printf("Error : FDCT : not enough memory for working vector.\n");
            error_exit();
        }
    }
```

```
    y = yt[logm-1];

    /* Copy x into y, with data shuffling */
    dx = 2;
    for (gr = 0; gr < logm; gr++) {
        xp = x + (1 << gr); yp = y + m2;
        for (n = 0; n < m2; n++) {
            *yp++ = *xp;
            xp += dx;
        }
        m2 >>= 1;
        dx <<= 1;
    }
    y[0]   = x[0];
    y[1] = x[m/2];

    /* Call recursive module */
    idctrec(y, logm);

    /* Copy y into x, with appropriate inverse scaling */
    tmp = 1.0 / m;
    xp = x; yp = y;
    for (n = 0; n < m; n++) {
        *xp++ = tmp * *yp++;
    }
}
```

Appendix C

Programs for Lapped Transforms

In this Appendix we present computer programs that implement the fast algorithms for the LOT (discussed in Section 4.3) and for the ELT (discussed in Section 5.4). All programs were written in ANSI-standard "C", and they should be able to run unmodified in a wide variety of machines. They have been extensively tested, and should produce correct results. However, the author does not warrant that the programs are absolutely free from errors or that they will meet any specific requirements. The responsibility for the use of information obtained from those programs lies entirely on the user.

The LOT module implements the type-II LOT, which is valid for any block length. It is based on the fast LOT flowgraph of Section 4.3, and so it calls the DCT and DCT-IV modules described in Appendix B. The ELT module is based on the flowgraphs presented in Section 5.4, and calls the DCT-IV module of Appendix B.

As we have seen in Sections 4.3 and 5.4, causal versions of the LOT and ELT need several z^{-1} and z^{-2} storage units. In the LOT and ELT modules described below, the storage for these delays is allocated inside the modules, and thus they are transparent to the user. In applications where ELTs of different signals should be computed within the same program, we need different buffers for each signal. Therefore, we have also included versions of the LOT and ELT module in which a pointer to the z^{-1} and z^{-2} storage locations is passed as an extra argument.

The ELT module gets the butterfly angles from the file angelt.txt. An example of such a file is presented at the end of this Appendix. Tables of butterfly angles are presented in Appendix D.

```
/*------------------------------------------------------------------*
 *    FLOT.C  -  Fast Lapped Orthogonal Transform                   *
 *                                                                  *
 *    This is the type-II LOT, as described in Section 4.4.         *
 *                                                                  *
 *    In order to minimize the number of multiplications, the       *
 *    transform is not orthogonal.  The basis functions have their  *
 *    energies equal to  1 / M.                                     *
 *                                                                  *
 *    Author:  Henrique S. Malvar.                                  *
 *    Date:    October 8, 1991.                                     *
 *                                                                  *
 *    Usage:   flot(x, logm);   -- direct transform                 *
 *             filot(x, logm);  -- inverse transform                *
 *                                                                  *
 *    Arguments:  x (float)   - input and output vector, length M.  *
 *                logm (int)  - log (base 2) of vector length M,    *
 *                              e.g., for M = 256 -> logm = 8.      *
 *                                                                  *
 *    Note: this is a causal version.  Thus, there is a delay of    *
 *    half a block in flot or filot.  Following a call to flot with *
 *    a call to filot will recover the original signal with one     *
 *    block delay.                                                  *
 *------------------------------------------------------------------*/

#include <stdio.h>
#include <stdlib.h>
#include <string.h>
#include <alloc.h>
#include <math.h>

#define  MAXLOGM    12    /* max LOT block size = 2^MAXLOGM */

void  fidct(float *, int);   /* fidct prototype */
void  fdct(float *, int);    /* fdct prototype */
void  fdctiv(float *, int);  /* fdctiv prototype */

/*------------------------------------------------------------------*
 *    Error exit for program abortion.                              *
 *------------------------------------------------------------------*/

static  void  error_exit(void)
{
   exit(1);
}
```

```
/*----------------------------------------------------------------------*
 *      Direct LOT                                                       *
 *----------------------------------------------------------------------*/

void flot(float *x, int logm)
{
    static    int      n, m, m2, m4;
    static    float    tmp, fac, *xp1, *xp2, *yp1, *yp2, *y;
    static    float    *yt[MAXLOGM];

    /* Check range of logm */
    if ((logm < 2) || (logm > MAXLOGM)) {
        printf("Error : FLOT : logm = %d is out of bounds [%d, %d]\n",
            logm, 2, MAXLOGM);
        error_exit();
    }

    /* Define m */
    m = 1 << logm;
    m2 = m / 2;
    m4 = m2 / 2;

    /* Allocate space for working vector, if necessary */
    if (yt[logm-2] == NULL) {
        if ((yt[logm-2] = (float *) calloc(3 * m2, sizeof(float))) == NULL) {
            printf("Error : FLOT : not enough memory for working vector.\n");
            error_exit();
        }
    }
    y = yt[logm-2];

    /* Compute DCT */
    fdct(x, logm);

    /* Copy even-indexed x's on y[0]  and
       odd-indexed  x's on y[m/2] */
    xp1 = x;
    yp1 = y;
    yp2 = y + m2;
    for (n = 0; n < m2; n++ ) {
        *(yp1++) = *(xp1++);
        *(yp2++) = *(xp1++);
    }

    /* First butterflies with +1/-1  factors, with  1/2  factor,
       output in  x   */
    xp1 = x;
    xp2 = x + m2;
    yp1 = y;
    yp2 = y + m2;
```

```
for (n = 0; n < m2; n++ ) {
    *(xp1++) = 0.5 * ( *yp1     + *yp2 );
    *(xp2++) = 0.5 * ( *(yp1++) - *(yp2++) );
}

/* This piece of code correspond to the  z^-1  delays in Section 4.4.
   The stored values are in the last  m/2  samples of  y  */
memcpy(y, y + m, m2 * sizeof(float));
memcpy(y + m, x + m2, m2 * sizeof(float));

/* Copy first  m/2  coefficients of  y  in  x[m/2]  */
memcpy(x + m2, y, m2 * sizeof(float));

/* Second stage of  +1/-1  butterflies, output in  y */
xp1 = x;
xp2 = x + m2;
yp1 = y;
yp2 = y + m2;
for (n = 0; n < m2; n++ ) {
    *(yp1++) = *xp1 + *xp2;
    *(yp2++) = *xp1 - *xp2;
    xp1++; xp2++;
}

/* Length-(n/2)  IDCT  */
fidct(y + m2, logm-1);

/* Length-(n/2)  DST-IV  via DCT-IV */
yp1 = y + m2 + 1;
for (n = 0; n < m4; n++) {
    *yp1 = - *yp1;
    yp1++; yp1++;
}
fdctiv(y + m2, logm-1);
yp1 = y + m2; yp2 = y + m - 1;
fac = sqrt(m2);
for (n = 0; n < m4; n++) {
    tmp  = *yp2;
    *yp2 = fac * *yp1;
    *yp1 = fac * tmp;
    yp1++; yp2--;
}

/* Even/odd re-indexing, output in  x  */
xp1 = x;
yp1 = y;
yp2 = y + m2;
```

```
   for (n = 0; n < m2; n++ ) {
      *(xp1++) = *(yp1++);
      *(xp1++) = *(yp2++);
   }
}
/*----------------------------------------------------------------------*
 *      Inverse LOT                                                      *
 *----------------------------------------------------------------------*/

void filot(float *x,  int logm)
{
   static   int     n, m, m2, m4;
   static   float   tmp, fac, *xp1, *xp2, *yp1, *yp2, *y;
   static   float   *yt[MAXLOGM];

   /* Check range of logm */
   if ((logm < 2) || (logm > MAXLOGM)) {
      printf("Error : FLOT : logm = %d is out of bounds [%d, %d]\n",
         logm, 2, MAXLOGM);
      error_exit();
   }

   /* Define m */
   m = 1 << logm;
   m2 = m / 2;
   m4 = m2 / 2;

   /* Allocate space for working vector, if necessary */
   if (yt[logm-2] == NULL) {
      if ((yt[logm-2] = (float *) calloc(3 * m2, sizeof(float))) == NULL) {
         printf("Error : FLOT : not enough memory for working vector.\n");
         error_exit();
      }
   }
   y = yt[logm-2];

   /* Even/odd re-indexing, output in  y  */
   xp1 = x;
   yp1 = y;
   yp2 = y + m2;
   for (n = 0; n < m2; n++ ) {
      *(yp1++) = *(xp1++);
      *(yp2++) = *(xp1++);
   }

   /* Length-(m/2)  DST-IV  */
   fac = sqrt(2.0 / m);
   yp1 = y + m2; yp2 = y + m - 1;
```

```
for (n = 0; n < m4; n++) {
   tmp  = *yp2;
   *yp2 = fac * *yp1;
   *yp1 = fac * tmp;
   yp1++; yp2--;
}
fdctiv(y + m2, logm-1);
yp1 = y + m2 + 1;
for (n = 0; n < m4; n++) {
   *yp1 = - *yp1;
   yp1++; yp1++;
}

/* Length-(n/2)  DCT */
fdct(y + m2, logm-1);

/* First butterflies with  +1/-1  factors, with  1/2  factor,
   output in  x */
xp1 = x;
xp2 = x + m2;
yp1 = y;
yp2 = y + m2;
for (n = 0; n < m2; n++ ) {
   *(xp1++) = 0.5 * ( *yp1     + *yp2 );
   *(xp2++) = 0.5 * ( *(yp1++) - *(yp2++) );
}

/* This piece of code correspond to the  z^(-1)  delays indicated in
   Section 4.4. The stored values are in the last  m/2  samples of y */
memcpy(y, y + m, m2 * sizeof(float));
memcpy(y + m, x, m2 * sizeof(float));

/* Copy first m/2  coefficients of  y  in  x[0] */
memcpy(x, y, m2 * sizeof(float));

/* Second stage of  +1/-1  butterflies, output in  y */
xp1 = x;
xp2 = x + m2;
yp1 = y;
yp2 = y + m2;
for (n = 0; n < m2; n++ ) {
   *(yp1++) = *xp1 + *xp2;
   *(yp2++) = *xp1 - *xp2;
   xp1++; xp2++;
}

/* Copy  y[0]  on  even-indexed  x's  and
   y[m/2]  on odd-indexed  x's */
xp1 = x;
yp1 = y;
```

```
    yp2 = y + m2;
    for (n = 0; n < m2; n++ ) {
      *(xp1++) = *(yp1++);
      *(xp1++) = *(yp2++);
    }

    /* Compute IDCT */
    fidct(x, logm);
}




/*----------------------------------------------------------------*
 *    FLOT2.C  -  Fast Lapped Orthogonal Transform                *
 *                                                                *
 *    This is the type-II LOT, as described in Section 4.4.       *
 *                                                                *
 *    Version 2: with buffer memory as an argument.               *
 *                                                                *
 *    Author:   Henrique S. Malvar.                               *
 *    Date:     October 8, 1991.                                  *
 *                                                                *
 *    Usage:    flot2(x, y, logm);   -- direct transform          *
 *              filot2(x, y, logm);  -- inverse transform         *
 *                                                                *
 *    Arguments:  x (float)   - input and output vector, length M.*
 *                y (float)   - pointer to internal buffers; it   *
 *                              should point to a previously      *
 *                              allocated buffer with room for    *
 *                              3 * M / 2  values.                *
 *                logm (int)  - log (base 2) of vector length M,  *
 *                              e.g., for M = 256 -> logm = 8.    *
 *----------------------------------------------------------------*/

#include <stdio.h>
#include <stdlib.h>
#include <string.h>
#include <alloc.h>
#include <math.h>

#define  MAXLOGM     12    /* max LOT block size = 2^MAXLOGM */

void  fidct(float *, int);    /* fidct prototype */
void  fdct(float *, int);     /* fdct prototype */
void  fdctiv(float *, int);   /* fdctiv prototype */
```

```
/*----------------------------------------------------------------------*
 *      Error exit for program abortion.                                *
 *----------------------------------------------------------------------*/

static   void  error_exit(void)
{
   exit(1);
}

/*----------------------------------------------------------------------*
 *      Direct LOT                                                      *
 *----------------------------------------------------------------------*/

void flot2(float *x, float *y, int logm)
{
   static   int     n, m, m2, m4;
   static   float   tmp, fac, *xp1, *xp2, *yp1, *yp2;

   /* Check range of logm */
   if ((logm < 2) || (logm > MAXLOGM)) {
      printf("Error : FLOT : logm = %d is out of bounds [%d, %d]\n",
         logm, 2, MAXLOGM);
      error_exit();
   }

   /* Define m */
   m = 1 << logm;
   m2 = m / 2;
   m4 = m2 / 2;

   /* Compute DCT */
   fdct(x, logm);

   /* Copy  even-indexed  x's  on  y[0]  and
      odd-indexed  x's on  y[m/2] */
   xp1 = x;
   yp1 = y;
   yp2 = y + m2;
   for (n = 0; n < m2; n++ ) {
      *(yp1++) = *(xp1++);
      *(yp2++) = *(xp1++);
   }

   /* First butterflies with  +1/-1  factors, with   1/2  factor,
      output in  x  */
   xp1 = x;
   xp2 = x + m2;
   yp1 = y;
   yp2 = y + m2;
```

```
for (n = 0; n < m2; n++ ) {
   *(xp1++) = 0.5 * ( *yp1     + *yp2 );
   *(xp2++) = 0.5 * ( *(yp1++) - *(yp2++) );
}

/* This piece of code correspond to the  z^-1  delays in Section 4.4.
   The stored values are in the last  m/2  samples of  y  */
memcpy(y, y + m, m2 * sizeof(float));
memcpy(y + m, x + m2, m2 * sizeof(float));

/* Copy first  m/2  coefficients of  y  in  x[m/2]  */
memcpy(x + m2, y, m2 * sizeof(float));

/* Second stage of  +1/-1  butterflies, output in  y */
xp1 = x;
xp2 = x + m2;
yp1 = y;
yp2 = y + m2;
for (n = 0; n < m2; n++ ) {
   *(yp1++) = *xp1 + *xp2;
   *(yp2++) = *xp1 - *xp2;
   xp1++; xp2++;
}

/* Length-(n/2)  IDCT  */
fidct(y + m2, logm-1);

/* Length-(n/2)  DST-IV  via DCT-IV */
yp1 = y + m2 + 1;
for (n = 0; n < m4; n++) {
   *yp1 = - *yp1;
   yp1++; yp1++;
}
fdctiv(y + m2, logm-1);
yp1 = y + m2; yp2 = y + m - 1;
fac = sqrt(m2);
for (n = 0; n < m4; n++) {
   tmp  = *yp2;
   *yp2 = fac * *yp1;
   *yp1 = fac * tmp;
   yp1++; yp2--;
}

/* Even/odd re-indexing, output in  x  */
xp1 = x;
yp1 = y;
yp2 = y + m2;
```

```
      for (n = 0; n < m2; n++ ) {
         *(xp1++) = *(yp1++);
         *(xp1++) = *(yp2++);
      }
}

/*-----------------------------------------------------------------------*
 *      Inverse LOT                                                      *
 *-----------------------------------------------------------------------*/

void filot2(float *x, float *y, int logm)
{
   static   int      n, m, m2, m4;
   static   float    tmp, fac, *xp1, *xp2, *yp1, *yp2;

   /* Check range of logm */
   if ((logm < 2) || (logm > MAXLOGM)) {
      printf("Error : FLOT : logm = %d is out of bounds [%d, %d]\n",
         logm, 2, MAXLOGM);
      error_exit();
   }

   /* Define m */
   m = 1 << logm;
   m2 = m / 2;
   m4 = m2 / 2;

   /* Even/odd re-indexing, output in  y  */
   xp1 = x;
   yp1 = y;
   yp2 = y + m2;
   for (n = 0; n < m2; n++ ) {
      *(yp1++) = *(xp1++);
      *(yp2++) = *(xp1++);
   }

   /* Length-(m/2)  DST-IV  */
   fac = sqrt(2.0 / m);
   yp1 = y + m2; yp2 = y + m - 1;
   for (n = 0; n < m4; n++) {
      tmp  = *yp2;
      *yp2 = fac * *yp1;
      *yp1 = fac * tmp;
      yp1++; yp2--;
   }
   fdctiv(y + m2, logm-1);
   yp1 = y + m2 + 1;
```

```
for (n = 0; n < m4; n++) {
   *yp1 = - *yp1;
   yp1++; yp1++;
}

/* Length-(n/2)  DCT */
fdct(y + m2, logm-1);

/* First butterflies with +1/-1  factors, with  1/2  factor,
   output in  x  */
xp1 = x;
xp2 = x + m2;
yp1 = y;
yp2 = y + m2;
for (n = 0; n < m2; n++ ) {
   *(xp1++) = 0.5 * ( *yp1     + *yp2 );
   *(xp2++) = 0.5 * ( *(yp1++) - *(yp2++) );
}

/* This piece of code correspond to the  z^(-1)  delays indicated in
   Section 4.4. The stored values are in the last  m/2  samples of y */
memcpy(y, y + m, m2 * sizeof(float));
memcpy(y + m, x, m2 * sizeof(float));

/* Copy first  m/2  coefficients of  y  in  x[0]  */
memcpy(x, y, m2 * sizeof(float));

/* Second stage of  +1/-1  butterflies, output in  y */
xp1 = x;
xp2 = x + m2;
yp1 = y;
yp2 = y + m2;
for (n = 0; n < m2; n++ ) {
   *(yp1++) = *xp1 + *xp2;
   *(yp2++) = *xp1 - *xp2;
   xp1++; xp2++;
}

/* Copy  y[0]  on  even-indexed  x's  and
   y[m/2]  on odd-indexed  x's */
xp1 = x;
yp1 = y;
yp2 = y + m2;
for (n = 0; n < m2; n++ ) {
   *(xp1++) = *(yp1++);
   *(xp1++) = *(yp2++);
}
```

```
    /* Compute IDCT */
    fidct(x, logm);
}
```

```
/*---------------------------------------------------------------*
 *    FELT.C  -  Fast Extended Lapped Transform                  *
 *                                                               *
 *    Algorithm: as described in  Section 5.4.   The MLT can also be  *
 *    computed with this module, by setting k = 1.               *
 *                                                               *
 *    This is an orthogonal transform.   All basis functions have  *
 *    unity energy.                                              *
 *                                                               *
 *    Author:   Henrique S. Malvar.                             *
 *    Date:     October 8, 1991.                                *
 *                                                               *
 *    Usage:    felt(x, k, logm);   -- direct transform          *
 *              fielt(x, k, logm);  -- inverse transform         *
 *                                                               *
 *    Arguments:  x (float)   - input and output vector, length M.  *
 *                k (int)     - overlapping factor.             *
 *                logm (int)  - log (base 2) of vector length M,  *
 *                              e.g., for M = 256 -> logm = 8.   *
 *                                                               *
 *    Note: this is a causal version.  Thus, there is a delay of  *
 *    (2 * k - 1) M / 2 samples in either felt or fielt.         *
 *    Following a call to felt with a call to fielt will recover  *
 *    the original signal with a total delay of (2 * k - 1) blocks.  *
 *                                                               *
 *--------------- Structure of the table of cosines ------------------*
 *                                                               *
 *    tab[k-1][logm-1] : is a pointer to the beginning of the    *
 *    table.  For a given overlapping factor k, and number of    *
 *    subbands M = 2^logm, there are k stages of butterflies     *
 *    (D_0 to D_k-1).  Each stage has m/2 butterflies, and each  *
 *    butterfly is of the form:                                 *
 *                                                               *
 *                    -c                                         *
 *        x_1 ------- x_1 <-- -c * x_1 + s * x_2                 *
 *             \  /s                                            *
 *              X          c = cos(ang), s = sin(ang)           *
 *             /  \s                                            *
 *        x_2 ------- x_2 <--  s * x_1 + c * x_2                 *
 *                    c                                         *
 *                                                               *
 *    Each butterfly can be computed with three multiplies and   *
 *    three adds, in the form                                    *
```

```
*                                                               *
*                     -(s+c)                                    *
*          x_1     --------- x_1                                *
*                    \  s  /                                    *
*                     -------                                   *
*                    /  c-s  \                                  *
*          x_2     --------- x_2                                *
*                                                               *
*    Thus, for each butterfly, we need three table entries, for: *
*    -(s+c), s, and (c-s). Because there are M/2 butterflies per *
*    stage, tab[k-1][logm-1] points to 3 * k * M/2 elements,    *
*    arranged in the form: the M/2 values of s [sin(ang)] for the *
*    M/2 butterflies of D_0, starting at the outermost butterfly; *
*    then, the M/2 corresponding values of -(s+c), and the M/2  *
*    corresponding values of (c-s). Then, the M/2 values of s for *
*    D_1, and so on. The last M/2 elements are the (c-s) values *
*    for D_k-1.                                                 *
*                                                               *
*--------- Structure of the buffers for the delay elements ----------*
*                                                               *
*    After each butterfly in the direct ELT, M/2 elements must go *
*    through a delay of z^-2, i.e., a delay of two blocks. After *
*    the last stage, the delay is only z^-1. Thus, for each stage *
*    but the last we need 2 * M/2 storage locations, and M/2    *
*    locations for the last stage. We need also M/2 temporary   *
*    storage locations due to the swapping of the top and bottom *
*    halves of the vector just before the DCT-IV.              *
*    Therefore, we need k * M locations, and that is the size of the *
*    area pointed by yt[k-1][logm-1]. The first M locations are *
*    for the z^-2 delay after D_k-1, the next M locations are for *
*    the z^-2 delay after D_k-2, and so on. The last M locations *
*    have two purposes: the top M/2 locations are for the z^-1  *
*    delay after D_0, and the bottom M/2 locations are for temporary *
*    storage, pointed by wk.                                    *
*                                                               *
*-------------------------------------------------------------*/

#include <stdio.h>
#include <stdlib.h>
#include <string.h>
#include <alloc.h>
#include <math.h>

#define  MAXLOGM     12    /* max ELT block size = 2^MAXLOGM */
#define  MAXK        4     /* max overlapping factor */
#define  PI          3.14159265358979323846

void  fdctiv(float *, int);   /* fdctiv prototype */
```

```
/*------------------------------------------------------------------------*
 *    Error exit for program abortion.                                    *
 *-----------------------------------------------------------------------*/

static   void  error_exit(void)
{
   exit(1);
}

/*------------------------------------------------------------------------*
 *    Module to read tables of butterfly coefficients                     *
 *-----------------------------------------------------------------------*/

static   float    *tab[MAXK][MAXLOGM];
#define  NCHARS    80

static   void  read_table(int k, int logm)
{
   int      n, m, m2, nel, gr;
   float    ang, co, si, *s, *spc, *cms;
   FILE     *af;
   char     txt[NCHARS], *tp;

   /* Define m */
   m = 1 << logm;
   m2 = m / 2;

   nel = 3 * k * m2;
   if ((tab[k-1][logm-1] = (float *) calloc(nel, sizeof(float))) == NULL) {
      printf("Error : FELT : not enough memory for butterfly coeffs.\n");
      error_exit();
   }

   /* Read angles into first m/2 elements of each group of 3*m/2
      elements, corresponding to each D_k */
   if ((af = fopen("angelt.txt", "r")) == NULL) {
      printf("Error : FELT : could not open file angelt.txt\n");
      error_exit();
   }

   /* Skip lines until the first line for the given  m  &  k */
   nel = 2 * m + 3 * logm + (k - 1) * m2;
   for (n = 1; n < nel; n++) {
      if (fgets(txt, NCHARS-1, af) == NULL) {
         printf("Error : FELT : error reading file angelt.txt\n");
         error_exit();
      }
   }
```

```
    /* Read angles */
    for (n = 0; n < m2; n++) {
        if (fgets(txt, NCHARS-1, af) == NULL) {
            printf("Error : FELT : error reading file angelt.txt\n");
            error_exit();
        }
        tp = strtok(txt, " ");
        s = tab[k-1][logm-1] + n;
        for (gr = 0; gr < k; gr++) {
            *s = atof(tp);
            tp = strtok(NULL, " ");
            s += 3 * m2;
        }
    }
    fclose(af);

    /* Compute sines and cosines */
    s = tab[k-1][logm-1]; spc = s + m2; cms = spc + m2;
    for (gr = 0; gr < k; gr++) {

        /* Compute tables */
        for (n = 0; n < m2; n++) {
            ang = *s * PI;
            co = cos(ang); si = sin(ang);
            *s++ = si; *spc++ = - (si + co); *cms++ = co - si;
        }
        s += m; spc += m; cms += m;
    }
}
/*-------------------------------------------------------------------------*
 *      Direct ELT                                                         *
 *-------------------------------------------------------------------------*/

void felt(float *x, int k, int logm)
{
    int     n, m, m2, nel, gr;
    float   tmp, *xp1, *xp2, *op1, *y, *yp, *wk;
    float   *s, *spc, *cms;
    static  float   *yt[MAXK][MAXLOGM];

    /* Check range of logm */
    if ((logm < 1) || (logm > MAXLOGM)) {
        printf("Error : FELT : logm = %d is out of bounds [%d, %d]\n",
            logm, 1, MAXLOGM);
        error_exit();
    }
```

```
/* Check range of k */
if ((k < 1) || (k > MAXK)) {
   printf("Error : FELT : k = %d is out of bounds [%d, %d]\n",
      k, 1, MAXK);
   error_exit();
}

/* Define m */
m = 1 << logm;
m2 = m / 2;

/* Compute table of butterfly coefficients, if necessary */
if (tab[k-1][logm-1] == NULL) read_table(k, logm);

/* Allocate space for working vector, if necessary */
if (yt[k-1][logm-1] == NULL) {
   nel = k * m;
   if ((yt[k-1][logm-1] = (float *) calloc(nel, sizeof(float)))
      == NULL) {
      printf("Error : FELT : not enough memory for working vector.\n");
      error_exit();
   }
}
y  = yt[k-1][logm-1];
wk = y + k * m - m2;

/* Compute groups of butterflies and delays */
yp = y;
s = tab[k-1][logm-1] + (k-1) * 3 * m2;
spc = s + m2; cms = spc + m2;
for (gr = k; gr > 0; gr--) {

   /* Compute butterfly */
   xp1 = x;
   xp2 = x + m - 1;
   op1 = wk;
   for (n = 0; n < m2; n++) {
      tmp = *s++ * (*xp1 + *xp2);
      *op1++ = *spc++ * *xp1 + tmp;
      *xp2   = *cms++ * *xp2 + tmp;
      xp1++; xp2--;
   }
   n = 2 * m;
   s -= n; spc -= n; cms -= n;

   /* Perform delay z^-2 or z^-1 in the last group */
   if (gr == 1) {
      memcpy(x, yp, m2 * sizeof(float));
      memcpy(yp, wk, m2 * sizeof(float));
```

```
        } else {
            memcpy(x, yp, m2 * sizeof(float));
            memcpy(yp, yp + m2, m2 * sizeof(float));
            memcpy(yp + m2, wk, m2 * sizeof(float));
        }
        yp += m;
    }

    /* Swap top and bottom halfs of  x */
    memcpy(wk, x, m2 * sizeof(float));
    memcpy(x, x + m2, m2 * sizeof(float));
    memcpy(x + m2, wk, m2 * sizeof(float));

    /* Compute DCT-IV */
    fdctiv(x, logm);
}
/*-------------------------------------------------------------------------*
 *      Inverse ELT                                                        *
 *-------------------------------------------------------------------------*/

void fielt(float *x, int k, int logm)
{
    int      n, m, m2, nel, gr;
    float    tmp, *xp1, *xp2, *op1, *y, *yp, *wk;
    float    *s, *spc, *cms;
    static   float   *yt[MAXK][MAXLOGM];

    /* Check range of logm */
    if ((logm < 1) || (logm > MAXLOGM)) {
        printf("Error : FELT : logm = %d is out of bounds [%d, %d]\n",
            logm, 1, MAXLOGM);
        error_exit();
    }

    /* Check range of k */
    if ((k < 1) || (k > MAXK)) {
        printf("Error : FELT : k = %d is out of bounds [%d, %d]\n",
            k, 1, MAXK);
        error_exit();
    }

    /* Define m */
    m = 1 << logm;
    m2 = m / 2;

    /* Compute table of butterfly coefficients, if necessary */
    if (tab[k-1][logm-1] == NULL) read_table(k, logm);
```

```
/* Allocate space for working vector, if necessary */
if (yt[k-1][logm-1] == NULL) {
   nel = k * m;
   if ((yt[k-1][logm-1] = (float *) calloc(nel, sizeof(float)))
       == NULL) {
       printf("Error : FELT : not enough memory for working vector.\n");
       error_exit();
   }
}
y  = yt[k-1][logm-1];
wk = y + k * m - m2;

/* Compute DCT-IV */
fdctiv(x, logm);

/* Swap top and bottom halves of  x */
memcpy(wk, x, m2 * sizeof(float));
memcpy(x, x + m2, m2 * sizeof(float));
memcpy(x + m2, wk, m2 * sizeof(float));

/* Compute groups of delays and butterflies */
yp = y + (k - 1) * m;
s = tab[k-1][logm-1]; spc = s + m2; cms = spc + m2;
for (gr = 0; gr < k; gr++) {

   /* Perform delay z^-2 or z^-1 in the first group */
   if (gr == 0) {
       memcpy(wk, yp, m2 * sizeof(float));
       memcpy(yp, x + m2, m2 * sizeof(float));
   } else {
       memcpy(wk, yp, m2 * sizeof(float));
       memcpy(yp, yp + m2, m2 * sizeof(float));
       memcpy(yp + m2, x + m2, m2 * sizeof(float));
   }
   yp -= m;

   /* Compute butterfly */
   xp1 = x;
   xp2 = x + m - 1;
   op1 = wk + m2 - 1;
   for (n = 0; n < m2; n++) {
       tmp = *s++ * (*xp1 + *op1);
       *xp1 = *spc++ * *xp1 + tmp;
       *xp2 = *cms++ * *op1 + tmp;
       xp1++; xp2--; op1--;
   }
   s += m; spc += m; cms += m;
}
}
```

```
/*------------------------------------------------------------------------*
 *      FELT2.C  -  Fast Extended Lapped Transform                        *
 *                                                                        *
 *      Version 2: with buffer memory as an argument.                     *
 *                                                                        *
 *      Algorithm: as described in  Section 5.4.  The MLT can also be     *
 *      computed with this module, by setting k = 1.                      *
 *                                                                        *
 *      This is an orthogonal transform.  All basis functions have        *
 *      unity energy.                                                     *
 *                                                                        *
 *      Author:  Henrique S. Malvar.                                      *
 *      Date:    October 8, 1991.                                         *
 *                                                                        *
 *      Usage:   felt2(x, y, k, logm);   -- direct transform             *
 *               fielt2(x, y, k, logm);  -- inverse transform            *
 *                                                                        *
 *      Arguments:  x (float)   - input and output vector, length m.     *
 *                  y (float)   - pointer to internal buffers; it         *
 *                                should point to a previously            *
 *                                allocated buffer with room for          *
 *                                k * M   values.                         *
 *                  k (int)     - overlapping factor.                     *
 *                  logm (int)  - log (base 2) of vector length M,        *
 *                                e.g., for M = 256 -> logm = 8.          *
 *                                                                        *
 *      See further comments in module felt.c.                           *
 *------------------------------------------------------------------------*/

#include <stdio.h>
#include <stdlib.h>
#include <string.h>
#include <alloc.h>
#include <math.h>

#define  MAXLOGM      12    /* max ELT block size = 2^MAXLOGM */
#define  MAXK         4     /* max overlapping factor */
#define  PI           3.14159265358979323846

void  fdctiv(float *, int);    /* fdctiv prototype */

/*------------------------------------------------------------------------*
 *      Error exit for program abortion.                                  *
 *------------------------------------------------------------------------*/

static   void  error_exit(void)
{
   exit(1);
}
```

```
/*---------------------------------------------------------------*
 *    Module to read tables of butterfly coefficients            *
 *---------------------------------------------------------------*/
static    float    *tab[MAXK][MAXLOGM];
#define  NCHARS    80

static    void  read_table(int k, int logm)
{
    int     n, m, m2, nel, gr;
    float   ang, co, si, *s, *spc, *cms;
    FILE    *af;
    char    txt[NCHARS], *tp;

    /* Define m */
    m = 1 << logm;
    m2 = m / 2;

    nel = 3 * k * m2;
    if ((tab[k-1][logm-1] = (float *) calloc(nel, sizeof(float))) == NULL) {
        printf("Error : FELT : not enough memory for butterfly coeffs.\n");
        error_exit();
    }

    /* Read angles into first m/2 elements of each group of 3*m/2
       elements, corresponding to each D_k */
    if ((af = fopen("angelt.txt", "r")) == NULL) {
        printf("Error : FELT : could not open file angelt.txt\n");
        error_exit();
    }

    /* Skip lines until the first line for the given  m  &  k */
    nel = 2 * m + 3 * logm + (k - 1) * m2;
    for (n = 1; n < nel; n++) {
        if (fgets(txt, NCHARS-1, af) == NULL) {
            printf("Error : FELT : error reading file angelt.txt\n");
            error_exit();
        }
    }

    /* Read angles */
    for (n = 0; n < m2; n++) {
        if (fgets(txt, NCHARS-1, af) == NULL) {
            printf("Error : FELT : error reading file angelt.txt\n");
            error_exit();
        }
        tp = strtok(txt, " ");
        s = tab[k-1][logm-1] + n;
```

```
        for (gr = 0; gr < k; gr++) {
            *s = atof(tp);
            tp = strtok(NULL, " ");
            s += 3 * m2;
        }
    }
    fclose(af);

    /* Compute sines and cosines */
    s = tab[k-1][logm-1]; spc = s + m2; cms = spc + m2;
    for (gr = 0; gr < k; gr++) {

        /* Compute tables */
        for (n = 0; n < m2; n++) {
            ang = *s * PI;
            co = cos(ang); si = sin(ang);
            *s++ = si; *spc++ = - (si + co); *cms++ = co - si;
        }
        s += m; spc += m; cms += m;
    }
}

/*----------------------------------------------------------------------*
 *      Direct ELT                                                      *
 *----------------------------------------------------------------------*/

void felt2(float *x, float *y, int k, int logm)
{
    int         n, m, m2, gr;
    float       tmp, *xp1, *xp2, *op1, *yp, *wk;
    float       *s, *spc, *cms;

    /* Check range of logm */
    if ((logm < 1) || (logm > MAXLOGM)) {
        printf("Error : FELT : logm = %d is out of bounds [%d, %d]\n",
            logm, 1, MAXLOGM);
        error_exit();
    }

    /* Check range of k */
    if ((k < 1) || (k > MAXK)) {
        printf("Error : FELT : k = %d is out of bounds [%d, %d]\n",
            k, 1, MAXK);
        error_exit();
    }

    /* Define m */
    m = 1 << logm;
    m2 = m / 2;
```

```c
/* Compute table of butterfly coefficients, if necessary */
if (tab[k-1][logm-1] == NULL) read_table(k, logm);

wk = y + k * m - m2;

/* Compute groups of butterflies and delays */
yp = y;
s = tab[k-1][logm-1] + (k-1) * 3 * m2;
spc = s + m2; cms = spc + m2;
for (gr = k; gr > 0; gr--) {

    /* Compute butterfly */
    xp1 = x;
    xp2 = x + m - 1;
    op1 = wk;
    for (n = 0; n < m2; n++) {
        tmp = *s++ * (*xp1 + *xp2);
        *op1++ = *spc++ * *xp1 + tmp;
        *xp2   = *cms++ * *xp2 + tmp;
        xp1++; xp2--;
    }
    n = 2 * m;
    s -= n; spc -= n; cms -= n;

    /* Perform delay z^-2 or z^-1 in the last group */
    if (gr == 1) {
        memcpy(x, yp, m2 * sizeof(float));
        memcpy(yp, wk, m2 * sizeof(float));
    } else {
        memcpy(x, yp, m2 * sizeof(float));
        memcpy(yp, yp + m2, m2 * sizeof(float));
        memcpy(yp + m2, wk, m2 * sizeof(float));
    }
    yp += m;
}

/* Swap top and bottom halfs of  x */
memcpy(wk, x, m2 * sizeof(float));
memcpy(x, x + m2, m2 * sizeof(float));
memcpy(x + m2, wk, m2 * sizeof(float));

/* Compute DCT-IV */
fdctiv(x, logm);
}
```

```
/*----------------------------------------------------------------------*
 *      Inverse ELT                                                     *
 *----------------------------------------------------------------------*/

void fielt2(float *x, float *y, int k, int logm)
{
    int       n, m, m2, gr;
    float     tmp, *xp1, *xp2, *op1, *yp, *wk;
    float     *s, *spc, *cms;

    /* Check range of logm */
    if ((logm < 1) || (logm > MAXLOGM)) {
        printf("Error : FELT : logm = %d is out of bounds [%d, %d]\n",
            logm, 1, MAXLOGM);
        error_exit();
    }

    /* Check range of k */
    if ((k < 1) || (k > MAXK)) {
        printf("Error : FELT : k = %d is out of bounds [%d, %d]\n",
            k, 1, MAXK);
        error_exit();
    }

    /* Define m */
    m = 1 << logm;
    m2 = m / 2;

    /* Compute table of butterfly coefficients, if necessary */
    if (tab[k-1][logm-1] == NULL) read_table(k, logm);

    wk = y + k * m - m2;

    /* Compute DCT-IV */
    fdctiv(x, logm);

    /* Swap top and bottom halfs of  x */
    memcpy(wk, x, m2 * sizeof(float));
    memcpy(x, x + m2, m2 * sizeof(float));
    memcpy(x + m2, wk, m2 * sizeof(float));

    /* Compute groups of delays and butterflies */
    yp = y + (k - 1) * m;
    s = tab[k-1][logm-1]; spc = s + m2; cms = spc + m2;
    for (gr = 0; gr < k; gr++) {

        /* Perform delay z^-2 or z^-1 in the first group */
        if (gr == 0) {
            memcpy(wk, yp, m2 * sizeof(float));
            memcpy(yp, x + m2, m2 * sizeof(float));
```

```
        } else {
            memcpy(wk, yp, m2 * sizeof(float));
            memcpy(yp, yp + m2, m2 * sizeof(float));
            memcpy(yp + m2, x + m2, m2 * sizeof(float));
        }
        yp -= m;

        /* Compute butterfly */
        xp1 = x;
        xp2 = x + m - 1;
        op1 = wk + m2 - 1;
        for (n = 0; n < m2; n++) {
            tmp = *s++ * (*xp1 + *op1);
            *xp1 = *spc++ * *xp1 + tmp;
            *xp2 = *cms++ * *op1 + tmp;
            xp1++; xp2--; op1--;
        }
        s += m; spc += m; cms += m;
    }
}
```

In the following list, we have an example of the file angelt.txt, which is read by
the fast ELT routines. We present here only the file corresponding to $\omega_s = 1.2\pi/M$.
Tables of ELT butterfly angles for other values of ω_s can be found in Appendix D.

```
-------------------------------------------
Butterfly Angles for the ELT, ws=1.2*pi/M
(c) 1991 Henrique S. Malvar
-------------------------------------------
M = 2
-------------------------------------------
    0.3642
    0.5259    0.6546
    0.4044    0.4501    0.4209
    0.4951    0.5923    0.5568    0.5845
-------------------------------------------
M = 4
-------------------------------------------
    0.4144
    0.3119
    0.5485    0.6138
    0.5117    0.7015
    0.4382    0.4328    0.4300
    0.3845    0.4784    0.4070
    0.5214    0.5933    0.5519    0.5421
    0.4811    0.5805    0.5421    0.6304
-------------------------------------------
M = 8
-------------------------------------------
    0.4352
    0.3935
    0.3417
    0.2817
    0.5619    0.5948
    0.5368    0.6340
    0.5187    0.6780
    0.5056    0.7256
    0.4463    0.4210    0.4412
    0.4352    0.4481    0.4170
    0.4173    0.4705    0.3957
    0.3497    0.4884    0.4216
    0.5273    0.5837    0.5589    0.5336
    0.5164    0.6019    0.5424    0.5503
    0.4980    0.5972    0.5361    0.5932
    0.4674    0.5651    0.5443    0.6656
```

```
------------------------------------------
M = 16
------------------------------------------
    0.4443
    0.4260
    0.4052
    0.3817
    0.3558
    0.3275
    0.2973
    0.2659
    0.5693    0.5858
    0.5549    0.6041
    0.5424    0.6237
    0.5317    0.6446
    0.5226    0.6666
    0.5150    0.6897
    0.5085    0.7134
    0.5028    0.7378
    0.4496    0.4143    0.4470
    0.4444    0.4291    0.4354
    0.4393    0.4425    0.4228
    0.4337    0.4548    0.4096
    0.4260    0.4659    0.3975
    0.4128    0.4760    0.3903
    0.3839    0.4849    0.3982
    0.3116    0.4925    0.4491
    0.5382    0.5888    0.5529    0.5168
    0.5346    0.6054    0.5420    0.5170
    0.5291    0.6194    0.5340    0.5208
    0.5223    0.6288    0.5282    0.5301
    0.5142    0.6301    0.5243    0.5483
    0.5042    0.6183    0.5228    0.5803
    0.4896    0.5872    0.5265    0.6320
    0.4489    0.5368    0.5565    0.7039
------------------------------------------
M = 32
------------------------------------------
    0.4486
    0.4401
    0.4309
    0.4211
    0.4107
    0.3996
    0.3879
    0.3755
    0.3625
    0.3490
    0.3348
```

0.3201			
0.3050			
0.2896			
0.2739			
0.2580			
0.5731	0.5814		
0.5655	0.5902		
0.5583	0.5994		
0.5516	0.6088		
0.5453	0.6187		
0.5395	0.6288		
0.5342	0.6393		
0.5293	0.6500		
0.5247	0.6610		
0.5206	0.6723		
0.5168	0.6838		
0.5133	0.6955		
0.5101	0.7074		
0.5070	0.7195		
0.5042	0.7317		
0.5014	0.7439		
0.4510	0.4107	0.4497	
0.4484	0.4183	0.4443	
0.4459	0.4257	0.4384	
0.4434	0.4327	0.4323	
0.4409	0.4395	0.4259	
0.4383	0.4459	0.4193	
0.4355	0.4520	0.4127	
0.4325	0.4579	0.4061	
0.4288	0.4634	0.3999	
0.4242	0.4687	0.3944	
0.4179	0.4737	0.3905	
0.4088	0.4784	0.3892	
0.3950	0.4829	0.3925	
0.3728	0.4871	0.4040	
0.3370	0.4909	0.4291	
0.2828	0.4941	0.4725	
0.5298	0.5746	0.5670	0.5299
0.5292	0.5817	0.5602	0.5308
0.5278	0.5884	0.5543	0.5324
0.5257	0.5945	0.5494	0.5347
0.5231	0.5998	0.5452	0.5383
0.5198	0.6037	0.5417	0.5434
0.5162	0.6060	0.5389	0.5502
0.5121	0.6068	0.5367	0.5589
0.5076	0.6057	0.5350	0.5696
0.5027	0.6027	0.5340	0.5823
0.4972	0.5977	0.5337	0.5972
0.4909	0.5906	0.5343	0.6144
0.4832	0.5812	0.5364	0.6340

0.4743	0.5703	0.5399	0.6551
0.4645	0.5606	0.5444	0.6751
0.4546	0.5537	0.5492	0.6924

M = 64

0.4507	
0.4466	
0.4423	
0.4378	
0.4333	
0.4285	
0.4236	
0.4186	
0.4134	
0.4080	
0.4025	
0.3968	
0.3909	
0.3849	
0.3787	
0.3724	
0.3659	
0.3592	
0.3524	
0.3455	
0.3384	
0.3312	
0.3239	
0.3164	
0.3089	
0.3012	
0.2935	
0.2857	
0.2778	
0.2699	
0.2620	
0.2540	
0.5751	0.5793
0.5712	0.5836
0.5674	0.5880
0.5637	0.5924
0.5601	0.5970
0.5566	0.6017
0.5532	0.6064
0.5500	0.6113
0.5469	0.6162
0.5438	0.6212
0.5409	0.6262

0.5382	0.6314	
0.5355	0.6366	
0.5329	0.6419	
0.5304	0.6473	
0.5281	0.6527	
0.5258	0.6582	
0.5236	0.6638	
0.5216	0.6694	
0.5196	0.6751	
0.5177	0.6809	
0.5159	0.6867	
0.5141	0.6926	
0.5125	0.6985	
0.5108	0.7045	
0.5093	0.7104	
0.5078	0.7165	
0.5063	0.7225	
0.5049	0.7286	
0.5035	0.7347	
0.5021	0.7408	
0.5007	0.7469	
0.4517	0.4087	0.4511
0.4504	0.4126	0.4484
0.4491	0.4165	0.4457
0.4478	0.4202	0.4428
0.4465	0.4239	0.4399
0.4453	0.4275	0.4369
0.4440	0.4310	0.4339
0.4428	0.4345	0.4307
0.4415	0.4379	0.4275
0.4403	0.4411	0.4243
0.4390	0.4444	0.4210
0.4377	0.4475	0.4176
0.4363	0.4506	0.4143
0.4349	0.4536	0.4110
0.4334	0.4565	0.4076
0.4317	0.4593	0.4044
0.4299	0.4621	0.4013
0.4279	0.4648	0.3983
0.4256	0.4674	0.3956
0.4230	0.4700	0.3932
0.4198	0.4725	0.3912
0.4161	0.4749	0.3897
0.4116	0.4773	0.3890
0.4061	0.4796	0.3893
0.3993	0.4819	0.3909
0.3907	0.4840	0.3942
0.3796	0.4861	0.3999
0.3657	0.4881	0.4085

0.3480	0.4900	0.4208	
0.3255	0.4918	0.4380	
0.2983	0.4934	0.4598	
0.2664	0.4949	0.4863	
0.5297	0.5727	0.5693	0.5301
0.5292	0.5759	0.5660	0.5309
0.5286	0.5791	0.5628	0.5317
0.5279	0.5822	0.5598	0.5326
0.5271	0.5853	0.5570	0.5337
0.5261	0.5882	0.5544	0.5350
0.5250	0.5910	0.5520	0.5364
0.5238	0.5937	0.5497	0.5381
0.5224	0.5961	0.5476	0.5400
0.5210	0.5983	0.5457	0.5423
0.5193	0.6001	0.5440	0.5450
0.5176	0.6016	0.5424	0.5480
0.5158	0.6027	0.5410	0.5515
0.5139	0.6035	0.5397	0.5554
0.5119	0.6038	0.5386	0.5597
0.5097	0.6038	0.5377	0.5645
0.5074	0.6032	0.5369	0.5699
0.5051	0.6022	0.5362	0.5757
0.5026	0.6007	0.5357	0.5821
0.4999	0.5987	0.5354	0.5890
0.4971	0.5961	0.5353	0.5965
0.4941	0.5931	0.5354	0.6045
0.4909	0.5895	0.5358	0.6131
0.4873	0.5855	0.5365	0.6222
0.4836	0.5810	0.5375	0.6317
0.4795	0.5763	0.5389	0.6415
0.4751	0.5715	0.5406	0.6515
0.4704	0.5668	0.5426	0.6613
0.4657	0.5626	0.5447	0.6707
0.4610	0.5589	0.5468	0.6795
0.4563	0.5558	0.5488	0.6878
0.4517	0.5532	0.5509	0.6955

Appendix D

Tables of ELT Butterfly Angles

In this Appendix we present three tables of butterfly angles for the fast ELT structures in Fig. 5.10. Each table corresponds to a particular choice of the stopband frequency ω_s, as discussed in Section 5.3. The angles were obtained by the ELT design algorithm, which was also presented in Section 5.3. The number of subbands M was varied form 2 to 64, and the overlapping factor K from 1 to 4. It is unlikely that some application will call for values of M and K outside of those ranges. We recall that M must be a power of two, because the fast ELT algorithm uses a length-M DCT-IV, which is based on a length-$M/2$ complex FFT [see Sections 2.5 and 5.4].

With the tables, the reader can prepare text files in the same format as that of the file angelt.txt of Appendix C. Such files would be read by the fast ELT routines also presented in Appendix C. As for the choice of ω_s, we recall from Section 5.4 that ω_s controls the shape of the subband frequency responses. Lower values of ω_s lead to narrower transition widths, whereas higher values of ω_s lead to higher stopband attenuations. Examples of the effect of ω_s in the subband frequency responses were presented in Figs. 5.5 to 5.7.

M	K=1	K=2		K=3			K=4			
2	0.3187	0.5563	0.6822	0.3647	0.4245	0.4434	0.4581	0.5964	0.6131	0.6003
4	0.3522	0.5864	0.6494	0.4038	0.3992	0.4335	0.4922	0.5922	0.6098	0.5727
	0.2848	0.5279	0.7160	0.3144	0.4490	0.4650	0.4304	0.6016	0.6105	0.6270
8	0.3683	0.6019	0.6334	0.4193	0.3855	0.4325	0.5137	0.5915	0.6037	0.5575
	0.3358	0.5714	0.6658	0.3865	0.4129	0.4366	0.4740	0.5942	0.6127	0.5867
	0.3020	0.5422	0.6991	0.3423	0.4379	0.4518	0.4438	0.5993	0.6124	0.6134
	0.2674	0.5139	0.7330	0.2834	0.4595	0.4813	0.4176	0.6036	0.6080	0.6408
16	0.3763	0.6098	0.6255	0.4211	0.3780	0.4374	0.5260	0.5919	0.5992	0.5491
	0.3604	0.5942	0.6414	0.4074	0.3921	0.4375	0.5025	0.5918	0.6072	0.5652
	0.3441	0.5789	0.6576	0.3961	0.4057	0.4344	0.4828	0.5932	0.6116	0.5797
	0.3275	0.5640	0.6740	0.3789	0.4189	0.4371	0.4657	0.5954	0.6133	0.5935
	0.3106	0.5494	0.6907	0.3544	0.4314	0.4470	0.4507	0.5980	0.6131	0.6068
	0.2934	0.5351	0.7076	0.3282	0.4431	0.4587	0.4369	0.6005	0.6116	0.6202
	0.2761	0.5209	0.7245	0.3074	0.4539	0.4650	0.4239	0.6027	0.6093	0.6338
	0.2587	0.5070	0.7415	0.2788	0.4640	0.4786	0.4112	0.6045	0.6066	0.6478
32	0.3802	0.6138	0.6216	0.4268	0.3749	0.4360	0.5332	0.5929	0.5960	0.5441
	0.3723	0.6059	0.6295	0.4145	0.3822	0.4416	0.5205	0.5923	0.6009	0.5527
	0.3644	0.5981	0.6374	0.4026	0.3894	0.4465	0.5087	0.5921	0.6050	0.5608
	0.3563	0.5903	0.6454	0.3916	0.3965	0.4505	0.4977	0.5924	0.6082	0.5686
	0.3482	0.5827	0.6535	0.3817	0.4034	0.4531	0.4877	0.5930	0.6105	0.5760
	0.3400	0.5751	0.6617	0.3734	0.4101	0.4541	0.4785	0.5938	0.6121	0.5831
	0.3317	0.5677	0.6699	0.3670	0.4166	0.4531	0.4699	0.5949	0.6130	0.5900
	0.3233	0.5603	0.6782	0.3624	0.4230	0.4502	0.4619	0.5961	0.6133	0.5967
	0.3148	0.5530	0.6865	0.3599	0.4293	0.4453	0.4544	0.5974	0.6132	0.6034
	0.3063	0.5458	0.6949	0.3593	0.4356	0.4384	0.4472	0.5987	0.6128	0.6101
	0.2977	0.5386	0.7033	0.3603	0.4420	0.4300	0.4403	0.5999	0.6120	0.6168
	0.2891	0.5315	0.7118	0.3620	0.4485	0.4209	0.4337	0.6011	0.6111	0.6235
	0.2805	0.5245	0.7203	0.3615	0.4551	0.4140	0.4271	0.6022	0.6099	0.6303
	0.2718	0.5174	0.7287	0.3574	0.4613	0.4108	0.4207	0.6032	0.6087	0.6372
	0.2631	0.5105	0.7372	0.3370	0.4663	0.4239	0.4144	0.6041	0.6073	0.6442
	0.2544	0.5035	0.7458	0.2860	0.4696	0.4676	0.4081	0.6049	0.6059	0.6514

Table D.1. Butterfly angles (fractions of π) for the fast ELT structure, $\omega_s = 0.8\pi/M$.

M	K=1	K=2		K=3			K=4			
64	0.3822	0.6158	0.6197	0.4291	0.3762	0.4316	0.5402	0.5957	0.5921	0.5407
	0.3783	0.6118	0.6236	0.4248	0.3797	0.4318	0.5378	0.5992	0.5904	0.5410
	0.3743	0.6079	0.6275	0.4208	0.3832	0.4318	0.5351	0.6024	0.5891	0.5416
	0.3703	0.6039	0.6314	0.4169	0.3866	0.4317	0.5323	0.6052	0.5882	0.5423
	0.3664	0.6000	0.6354	0.4132	0.3900	0.4315	0.5292	0.6076	0.5877	0.5434
	0.3624	0.5961	0.6394	0.4097	0.3933	0.4311	0.5259	0.6097	0.5875	0.5448
	0.3583	0.5923	0.6434	0.4065	0.3966	0.4304	0.5225	0.6116	0.5876	0.5464
	0.3543	0.5884	0.6474	0.4036	0.3999	0.4295	0.5190	0.6135	0.5878	0.5480
	0.3502	0.5846	0.6515	0.4009	0.4032	0.4284	0.5152	0.6152	0.5881	0.5499
	0.3461	0.5808	0.6555	0.3985	0.4065	0.4271	0.5113	0.6166	0.5885	0.5522
	0.3420	0.5770	0.6596	0.3965	0.4097	0.4255	0.5071	0.6177	0.5890	0.5549
	0.3379	0.5733	0.6637	0.3946	0.4129	0.4236	0.5028	0.6187	0.5894	0.5579
	0.3337	0.5695	0.6678	0.3932	0.4161	0.4215	0.4984	0.6195	0.5897	0.5612
	0.3296	0.5658	0.6720	0.3921	0.4193	0.4191	0.4939	0.6201	0.5899	0.5647
	0.3254	0.5621	0.6761	0.3913	0.4225	0.4163	0.4893	0.6204	0.5900	0.5686
	0.3212	0.5585	0.6803	0.3908	0.4257	0.4133	0.4847	0.6206	0.5901	0.5727
	0.3169	0.5548	0.6844	0.3905	0.4289	0.4101	0.4800	0.6205	0.5904	0.5770
	0.3127	0.5512	0.6886	0.3905	0.4321	0.4067	0.4752	0.6201	0.5909	0.5815
	0.3085	0.5476	0.6928	0.3907	0.4353	0.4031	0.4703	0.6196	0.5916	0.5861
	0.3042	0.5440	0.6970	0.3910	0.4386	0.3994	0.4654	0.6190	0.5925	0.5907
	0.2999	0.5404	0.7012	0.3915	0.4418	0.3956	0.4603	0.6180	0.5937	0.5955
	0.2956	0.5368	0.7054	0.3919	0.4450	0.3918	0.4553	0.6169	0.5950	0.6003
	0.2913	0.5333	0.7097	0.3922	0.4482	0.3883	0.4504	0.6155	0.5963	0.6053
	0.2870	0.5297	0.7139	0.3920	0.4512	0.3852	0.4455	0.6137	0.5975	0.6106
	0.2827	0.5262	0.7181	0.3910	0.4542	0.3829	0.4407	0.6115	0.5986	0.6165
	0.2783	0.5227	0.7224	0.3889	0.4569	0.3818	0.4359	0.6089	0.5995	0.6230
	0.2740	0.5192	0.7266	0.3843	0.4592	0.3832	0.4309	0.6064	0.6005	0.6297
	0.2696	0.5157	0.7309	0.3750	0.4608	0.3892	0.4254	0.6048	0.6023	0.6362
	0.2653	0.5122	0.7351	0.3611	0.4616	0.3999	0.4192	0.6048	0.6053	0.6416
	0.2609	0.5087	0.7394	0.3447	0.4622	0.4131	0.4134	0.6064	0.6083	0.6455
	0.2566	0.5052	0.7436	0.3232	0.4623	0.4314	0.4086	0.6087	0.6105	0.6487
	0.2522	0.5017	0.7479	0.2555	0.4605	0.4959	0.4067	0.6091	0.6086	0.6522

Table D.1. Continued.

M	K=1	K=2		K=3			K=4			
2	0.3436	0.5382	0.6676	0.3903	0.4416	0.4284	0.4816	0.5902	0.5774	0.5938
4	0.3875	0.5635	0.6294	0.4246	0.4187	0.4298	0.5103	0.5911	0.5766	0.5580
	0.2989	0.5181	0.7085	0.3484	0.4686	0.4376	0.4566	0.5841	0.5739	0.6341
8	0.4072	0.5779	0.6111	0.4352	0.4053	0.4363	0.5257	0.5909	0.5758	0.5413
	0.3675	0.5505	0.6484	0.4153	0.4336	0.4236	0.5040	0.5993	0.5683	0.5657
	0.3229	0.5280	0.6881	0.3816	0.4586	0.4226	0.4767	0.5939	0.5674	0.6060
	0.2747	0.5089	0.7292	0.3082	0.4794	0.4600	0.4439	0.5799	0.5732	0.6560
16	0.4164	0.5856	0.6023	0.4396	0.3981	0.4399	0.5396	0.5973	0.5690	0.5253
	0.3979	0.5706	0.6202	0.4311	0.4130	0.4327	0.5311	0.6077	0.5628	0.5308
	0.3780	0.5569	0.6389	0.4214	0.4271	0.4261	0.5214	0.6147	0.5581	0.5404
	0.3569	0.5445	0.6582	0.4094	0.4405	0.4212	0.5102	0.6174	0.5551	0.5547
	0.3345	0.5333	0.6780	0.3932	0.4530	0.4200	0.4972	0.6146	0.5541	0.5750
	0.3112	0.5230	0.6983	0.3695	0.4646	0.4260	0.4808	0.6049	0.5566	0.6027
	0.2870	0.5135	0.7188	0.3333	0.4750	0.4441	0.4595	0.5886	0.5639	0.6374
	0.2624	0.5044	0.7396	0.2808	0.4839	0.4783	0.4348	0.5722	0.5751	0.6726
32	0.4208	0.5896	0.5979	0.4453	0.3990	0.4435	0.5348	0.5856	0.5759	0.5325
	0.4119	0.5818	0.6067	0.4434	0.4067	0.4376	0.5352	0.5958	0.5683	0.5299
	0.4027	0.5742	0.6157	0.4418	0.4143	0.4311	0.5340	0.6051	0.5624	0.5284
	0.3931	0.5671	0.6248	0.4405	0.4216	0.4242	0.5316	0.6134	0.5578	0.5281
	0.3832	0.5602	0.6342	0.4395	0.4288	0.4168	0.5283	0.6206	0.5541	0.5290
	0.3729	0.5537	0.6437	0.4386	0.4357	0.4090	0.5244	0.6266	0.5510	0.5315
	0.3623	0.5475	0.6533	0.4379	0.4424	0.4010	0.5202	0.6312	0.5484	0.5354
	0.3514	0.5416	0.6631	0.4371	0.4489	0.3929	0.5155	0.6343	0.5462	0.5409
	0.3402	0.5360	0.6730	0.4364	0.4552	0.3845	0.5104	0.6355	0.5445	0.5484
	0.3288	0.5306	0.6830	0.4353	0.4613	0.3764	0.5044	0.6346	0.5437	0.5582
	0.3171	0.5255	0.6931	0.4333	0.4673	0.3691	0.4973	0.6309	0.5440	0.5709
	0.3052	0.5206	0.7034	0.4296	0.4730	0.3634	0.4889	0.6235	0.5454	0.5874
	0.2931	0.5158	0.7137	0.4213	0.4784	0.3623	0.4791	0.6115	0.5482	0.6087
	0.2809	0.5112	0.7240	0.4083	0.4836	0.3658	0.4662	0.5959	0.5539	0.6337
	0.2686	0.5067	0.7344	0.3763	0.4883	0.3882	0.4505	0.5819	0.5627	0.6573
	0.2562	0.5022	0.7448	0.3035	0.4921	0.4513	0.4353	0.5718	0.5712	0.6770

Table D.2. Butterfly angles (fractions of π) for the fast ELT structure, $\omega_s = \pi/M$.

M	K=1	K=2		K=3			K=4			
64	0.4230	0.5916	0.5958	0.4459	0.3970	0.4448	0.5348	0.5836	0.5775	0.5323
	0.4186	0.5876	0.6001	0.4450	0.4009	0.4419	0.5355	0.5890	0.5731	0.5306
	0.4142	0.5837	0.6045	0.4439	0.4048	0.4390	0.5359	0.5944	0.5692	0.5290
	0.4097	0.5799	0.6089	0.4430	0.4086	0.4359	0.5359	0.5996	0.5656	0.5276
	0.4050	0.5761	0.6134	0.4423	0.4124	0.4326	0.5355	0.6046	0.5625	0.5266
	0.4003	0.5724	0.6179	0.4417	0.4161	0.4292	0.5347	0.6092	0.5598	0.5259
	0.3955	0.5689	0.6225	0.4412	0.4198	0.4255	0.5337	0.6137	0.5573	0.5254
	0.3907	0.5653	0.6271	0.4404	0.4234	0.4222	0.5325	0.6182	0.5550	0.5250
	0.3857	0.5619	0.6318	0.4398	0.4269	0.4185	0.5313	0.6225	0.5528	0.5248
	0.3806	0.5586	0.6365	0.4392	0.4305	0.4149	0.5297	0.6263	0.5510	0.5251
	0.3755	0.5553	0.6413	0.4388	0.4339	0.4110	0.5278	0.6296	0.5494	0.5260
	0.3703	0.5521	0.6461	0.4383	0.4373	0.4072	0.5258	0.6327	0.5479	0.5271
	0.3650	0.5490	0.6509	0.4380	0.4407	0.4031	0.5237	0.6355	0.5466	0.5286
	0.3596	0.5460	0.6557	0.4378	0.4440	0.3989	0.5215	0.6379	0.5454	0.5304
	0.3542	0.5430	0.6606	0.4376	0.4473	0.3946	0.5191	0.6400	0.5444	0.5327
	0.3486	0.5402	0.6656	0.4374	0.4505	0.3903	0.5166	0.6416	0.5435	0.5354
	0.3431	0.5374	0.6705	0.4370	0.4536	0.3862	0.5140	0.6427	0.5427	0.5388
	0.3374	0.5346	0.6755	0.4366	0.4567	0.3821	0.5112	0.6428	0.5421	0.5430
	0.3317	0.5319	0.6805	0.4360	0.4598	0.3780	0.5083	0.6425	0.5416	0.5478
	0.3259	0.5293	0.6856	0.4352	0.4628	0.3742	0.5053	0.6418	0.5412	0.5530
	0.3200	0.5267	0.6906	0.4343	0.4657	0.3705	0.5021	0.6399	0.5410	0.5594
	0.3141	0.5242	0.6957	0.4331	0.4687	0.3670	0.4987	0.6374	0.5409	0.5663
	0.3082	0.5218	0.7008	0.4314	0.4715	0.3640	0.4951	0.6341	0.5411	0.5742
	0.3022	0.5193	0.7059	0.4289	0.4743	0.3617	0.4910	0.6297	0.5417	0.5832
	0.2962	0.5170	0.7111	0.4254	0.4771	0.3605	0.4864	0.6241	0.5429	0.5934
	0.2901	0.5146	0.7162	0.4206	0.4797	0.3606	0.4809	0.6171	0.5449	0.6051
	0.2840	0.5123	0.7214	0.4130	0.4823	0.3634	0.4743	0.6090	0.5478	0.6179
	0.2778	0.5101	0.7266	0.4004	0.4848	0.3712	0.4666	0.5998	0.5520	0.6319
	0.2717	0.5078	0.7318	0.3829	0.4871	0.3839	0.4579	0.5898	0.5571	0.6467
	0.2655	0.5055	0.7370	0.3629	0.4892	0.3991	0.4501	0.5817	0.5614	0.6596
	0.2593	0.5033	0.7422	0.3373	0.4912	0.4200	0.4435	0.5756	0.5646	0.6704
	0.2531	0.5011	0.7474	0.2563	0.4928	0.4961	0.4255	0.5620	0.5788	0.6893

Table D.2. Continued.

M	K=1	K=2		K=3			K=4			
2	0.3642	0.5259	0.6546	0.4044	0.4501	0.4209	0.4951	0.5923	0.5568	0.5845
4	0.4144	0.5485	0.6138	0.4382	0.4328	0.4300	0.5214	0.5933	0.5519	0.5421
	0.3119	0.5117	0.7015	0.3845	0.4784	0.4070	0.4811	0.5805	0.5421	0.6304
8	0.4352	0.5619	0.5948	0.4463	0.4210	0.4412	0.5273	0.5837	0.5589	0.5336
	0.3935	0.5368	0.6340	0.4352	0.4481	0.4170	0.5164	0.6019	0.5424	0.5503
	0.3417	0.5187	0.6780	0.4173	0.4705	0.3957	0.4980	0.5972	0.5361	0.5932
	0.2817	0.5056	0.7256	0.3497	0.4884	0.4216	0.4674	0.5651	0.5443	0.6656
16	0.4443	0.5693	0.5858	0.4496	0.4143	0.4470	0.5382	0.5888	0.5529	0.5168
	0.4260	0.5549	0.6041	0.4444	0.4291	0.4354	0.5346	0.6054	0.5420	0.5170
	0.4052	0.5424	0.6237	0.4393	0.4425	0.4228	0.5291	0.6194	0.5340	0.5208
	0.3817	0.5317	0.6446	0.4337	0.4548	0.4096	0.5223	0.6288	0.5282	0.5301
	0.3558	0.5226	0.6666	0.4260	0.4659	0.3975	0.5142	0.6301	0.5243	0.5483
	0.3275	0.5150	0.6897	0.4128	0.4760	0.3903	0.5042	0.6183	0.5228	0.5803
	0.2973	0.5085	0.7134	0.3839	0.4849	0.3982	0.4896	0.5872	0.5265	0.6320
	0.2659	0.5028	0.7378	0.3116	0.4925	0.4491	0.4489	0.5368	0.5565	0.7039
32	0.4486	0.5731	0.5814	0.4510	0.4107	0.4497	0.5298	0.5746	0.5670	0.5299
	0.4401	0.5655	0.5902	0.4484	0.4183	0.4443	0.5292	0.5817	0.5602	0.5308
	0.4309	0.5583	0.5994	0.4459	0.4257	0.4384	0.5278	0.5884	0.5543	0.5324
	0.4211	0.5516	0.6088	0.4434	0.4327	0.4323	0.5257	0.5945	0.5494	0.5347
	0.4107	0.5453	0.6187	0.4409	0.4395	0.4259	0.5231	0.5998	0.5452	0.5383
	0.3996	0.5395	0.6288	0.4383	0.4459	0.4193	0.5198	0.6037	0.5417	0.5434
	0.3879	0.5342	0.6393	0.4355	0.4520	0.4127	0.5162	0.6060	0.5389	0.5502
	0.3755	0.5293	0.6500	0.4325	0.4579	0.4061	0.5121	0.6068	0.5367	0.5589
	0.3625	0.5247	0.6610	0.4288	0.4634	0.3999	0.5076	0.6057	0.5350	0.5696
	0.3490	0.5206	0.6723	0.4242	0.4687	0.3944	0.5027	0.6027	0.5340	0.5823
	0.3348	0.5168	0.6838	0.4179	0.4737	0.3905	0.4972	0.5977	0.5337	0.5972
	0.3201	0.5133	0.6955	0.4088	0.4784	0.3892	0.4909	0.5906	0.5343	0.6144
	0.3050	0.5101	0.7074	0.3950	0.4829	0.3925	0.4832	0.5812	0.5364	0.6340
	0.2896	0.5070	0.7195	0.3728	0.4871	0.4040	0.4743	0.5703	0.5399	0.6551
	0.2739	0.5042	0.7317	0.3370	0.4909	0.4291	0.4645	0.5606	0.5444	0.6751
	0.2580	0.5014	0.7439	0.2828	0.4941	0.4725	0.4546	0.5537	0.5492	0.6924

Table D.3. Butterfly angles (fractions of π) for the fast ELT structure, $\omega_s = 1.2\pi/M$.

M	K=1	K=2		K=3			K=4			
64	0.4507	0.5751	0.5793	0.4517	0.4087	0.4511	0.5297	0.5727	0.5693	0.5301
	0.4466	0.5712	0.5836	0.4504	0.4126	0.4484	0.5292	0.5759	0.5660	0.5309
	0.4423	0.5674	0.5880	0.4491	0.4165	0.4457	0.5286	0.5791	0.5628	0.5317
	0.4378	0.5637	0.5924	0.4478	0.4202	0.4428	0.5279	0.5822	0.5598	0.5326
	0.4333	0.5601	0.5970	0.4465	0.4239	0.4399	0.5271	0.5853	0.5570	0.5337
	0.4285	0.5566	0.6017	0.4453	0.4275	0.4369	0.5261	0.5882	0.5544	0.5350
	0.4236	0.5532	0.6064	0.4440	0.4310	0.4339	0.5250	0.5910	0.5520	0.5364
	0.4186	0.5500	0.6113	0.4428	0.4345	0.4307	0.5238	0.5937	0.5497	0.5381
	0.4134	0.5469	0.6162	0.4415	0.4379	0.4275	0.5224	0.5961	0.5476	0.5400
	0.4080	0.5438	0.6212	0.4403	0.4411	0.4243	0.5210	0.5983	0.5457	0.5423
	0.4025	0.5409	0.6262	0.4390	0.4444	0.4210	0.5193	0.6001	0.5440	0.5450
	0.3968	0.5382	0.6314	0.4377	0.4475	0.4176	0.5176	0.6016	0.5424	0.5480
	0.3909	0.5355	0.6366	0.4363	0.4506	0.4143	0.5158	0.6027	0.5410	0.5515
	0.3849	0.5329	0.6419	0.4349	0.4536	0.4110	0.5139	0.6035	0.5397	0.5554
	0.3787	0.5304	0.6473	0.4334	0.4565	0.4076	0.5119	0.6038	0.5386	0.5597
	0.3724	0.5281	0.6527	0.4317	0.4593	0.4044	0.5097	0.6038	0.5377	0.5645
	0.3659	0.5258	0.6582	0.4299	0.4621	0.4013	0.5074	0.6032	0.5369	0.5699
	0.3592	0.5236	0.6638	0.4279	0.4648	0.3983	0.5051	0.6022	0.5362	0.5757
	0.3524	0.5216	0.6694	0.4256	0.4674	0.3956	0.5026	0.6007	0.5357	0.5821
	0.3455	0.5196	0.6751	0.4230	0.4700	0.3932	0.4999	0.5987	0.5354	0.5890
	0.3384	0.5177	0.6809	0.4198	0.4725	0.3912	0.4971	0.5961	0.5353	0.5965
	0.3312	0.5159	0.6867	0.4161	0.4749	0.3897	0.4941	0.5931	0.5354	0.6045
	0.3239	0.5141	0.6926	0.4116	0.4773	0.3890	0.4909	0.5895	0.5358	0.6131
	0.3164	0.5125	0.6985	0.4061	0.4796	0.3893	0.4873	0.5855	0.5365	0.6222
	0.3089	0.5108	0.7045	0.3993	0.4819	0.3909	0.4836	0.5810	0.5375	0.6317
	0.3012	0.5093	0.7104	0.3907	0.4840	0.3942	0.4795	0.5763	0.5389	0.6415
	0.2935	0.5078	0.7165	0.3796	0.4861	0.3999	0.4751	0.5715	0.5406	0.6515
	0.2857	0.5063	0.7225	0.3657	0.4881	0.4085	0.4704	0.5668	0.5426	0.6613
	0.2778	0.5049	0.7286	0.3480	0.4900	0.4208	0.4657	0.5626	0.5447	0.6707
	0.2699	0.5035	0.7347	0.3255	0.4918	0.4380	0.4610	0.5589	0.5468	0.6795
	0.2620	0.5021	0.7408	0.2983	0.4934	0.4598	0.4563	0.5558	0.5488	0.6878
	0.2540	0.5007	0.7469	0.2664	0.4949	0.4863	0.4517	0.5532	0.5509	0.6955

Table D.3. Continued.

Index

W

About the Author

Henrique Sarmento Malvar earned his Ph.D. degree from the Massachusetts Institute of Technology. Since 1979 he has been on the Faculty of the Universidade de Brasília, Brazil, where he is now Associate Professor of Electrical Engineering. Dr. Malvar is a senior member of the IEEE and a member of the Sigma Xi.

The Artech House Telecommunications Library

Vinton G. Cerf, *Series Editor*

Introduction to Telephones and Telephone Systems, Second Edition, by A. Michael Noll

Jitter Analysis in Digital System Design, by Yoshitaka Takasaki

Jitter in Digital Transmission Systems by Patrick R. Trischitta and Eve L. Varma

Implementing X.400 and X.500: The PP and QUIPU Systems, by Steve Kille

LANs to WANs: Network Management in the 1990s by Nathan J. Muller and Robert P. Davidson

Long Distance Services: A Buyer's Guide by Daniel D. Briere

Manager's Guide to CENTREX by John R. Abrahams

Measurement of Optical Fibers and Devices by G. Cancellieri and U. Ravaioli

Meteor Burst Communications by Jacob Z. Schanker

Minimum Risk Strategy for Acquiring Communications Equipment and Services by Nathan J. Muller

Mobile Information Systems by John Walker

Optical Fiber Transmission Systems by Siegried Geckeler

Optimization of Digital Transmission Systems by K. Trondle and G. Soder

Principles of Secure Communication Systems by Don J. Torrieri

Private Telecommunication Networks by Bruce Elbert

Radiodetermination Satellite Services and Standards by Martin Rothblatt

Residential Fiber Optic Networks: An Engineering and Economic Analysis by David P. Reed

Setting Global Telecommunication Standards: The Stakes, The Players, and The Process by Gerd Wallenstein

Techniques in Data Communications by Ralph Glasgal

Telecommunication Systems by Pierre-Girard Fontolliet

Telecommunications Technology Handbook by Daniel Minoli

Telephone Company and Cable Television Competition, Stuart N. Brotman, ed.

Television Technology: Fundamentals and Future Prospects by A. Michael Noll

Terrestrial Digital Microwave Communications, Ferdo Ivanek, ed.

The Telecommunications Deregulation Sourcebook, Stuart N. Brotman, ed.

Traffic Flow in Switching Systems by G. Hebuterne